Alan J. Anderson

3-6-67.

KING SOLOMON'S RING

KING SOLOMON'S RING

New Light on Animal Ways

KONRAD Z. LORENZ

Illustrated by the Author

Translated by
MARJORIE KERR WILSON

With a Foreword by
JULIAN HUXLEY

THE REPRINT SOCIETY LONDON

THIS TRANSLATION FROM THE GERMAN, BY MARJORIE KERR WILSON
FIRST PUBLISHED 1952
THIS EDITION FIRST PUBLISHED BY THE REPRINT SOCIETY LTD.
BY ARRANGEMENT WITH MESSRS. METHUEN AND CO. LTD.
1953

PRINTED IN GREAT BRITAIN BY RICHARD CLAY AND COMPANY, LTD.,
BUNGAY, SUFFOLK

To

MR AND MRS J. B. PRIESTLEY

without whose timely help
jackdaws would not — in
all probability — be flying
round Altenberg any more

CONTENTS

FOREWORD

by JULIAN HUXLEY

KONRAD LORENZ is one of the outstanding naturalists of our day. I have heard him referred to as the modern Fabre, but with birds and fishes instead of insects and spiders as his subject-matter. However, he is more than that, for he is not only, like Fabre, a provider of an enormous volume of new facts and penetrating observations, with a style of distinction and charm, but in addition has contributed in no small degree to the basic principles and theories of animal mind and behaviour. For instance, it is to him more than any other single man that we owe our knowledge of the existence of the strange biological phenomena of "releaser" and "imprinting" mechanisms.

The reader of this book who has followed the account of how Lorenz himself became "imprinted" on his baby goslings as their parent, or how his jackdaws regarded him as their general leader and companion, but chose other corvine birds (so long as they were on the wing) as flight companions, and fixed on his maidservant as a "love-object"; or how certain attitudes or gestures on the part of a fighting-fish or a wolf will act as releasers to promote or inhibit combat reactions in another individual of the species, will realize not only the strangeness of the facts but the fundamental nature of the principles that underlie them.

Of course, other naturalists too have worked along similar lines. I think of the pioneering studies of Lloyd Morgan in Britain, of Whitman in America, of the Heinroths in Germany; of the remarkable researches of the late Kingsley Noble of New York on the behaviour of lizards, and of Tinbergen of Holland and Oxford on releasers in sticklebacks and herring-gulls; and of the detailed illustration of the principles involved by a host of observers and students, most of them ornithologists, in western Europe and North America. But it remains true that Lorenz has done more than any single man to establish the principles and to formulate the essential ideas behind them. And then Lorenz has given himself over, body and soul, to his self-appointed task of really understanding animals, more thoroughly than any other biologist-naturalist that I can think of. This has involved keeping his objects of study in what amounts to the wild state, with full freedom of movement. His readers will discover all that this has meant in the way of hard work and inconvenience—sometimes amusing in retrospect, but usually awkward enough or even serious at the time.

But the labour and the inconvenience have been abundantly justified by the results. Indeed they were necessary, for thanks to such work by Lorenz (and by other devoted lovers and students of animals) it has become clear that animals do not reveal the higher possibilities of their nature and behaviour, nor the full range of their individual diversity, except in such conditions of freedom. Captivity cages minds as well as bodies, and rigid experimental procedure limits the range of performance; while freedom liberates the creatures' capacities and permits the observer to study their fullest developments.

The value of Lorenz's methods is strikingly exemplified in his long chapter on his jackdaws—one of the most illuminating accounts ever given of the life of a social organism. The strange blend of automatic reaction, intelligence and insight shown by these birds; the curious mechanisms of their social behaviour, which on the whole make for law and order and the safeguarding of weaker members of the colony (though none of the behaviour is undertaken with any such purpose in view); the difference between avian communication and human language; the presence of what, if it were to be exhibited by men, would be called chivalrous behaviour (but its total absence in non-social species like the turtle-dove, which in spite of its gentle reputation can be guilty of the most brutal cruelty to a defeated rival which cannot escape), the extraordinary and I believe the only established case of the social transmission of the knowledge that certain creatures are to be treated as enemies—all this and much else is set forth by Lorenz in such a way that his readers will never again be guilty of anthropomorphizing a bird, nor of the equal intellectual misdemeanour of "mechanomorphizing" it and reducing it to the false over-simplification of a mere system of reflexes.

However, it is not only with birds that Lorenz is at home. His account of the reproductive life of fighting-fish and sticklebacks—the combats and displays of the males; the reactions of the females, the males' parental care of their young—is equally brilliant and penetrating. If the behaviour of fish does not rise quite to the same height as that of birds, it is certainly much more extraordinary than most people have any idea of. And the description of how a certain male fighting-fish resolved a conflict is an admirable scientific account of a very unusual phenomenon—an animal making up its mind when it possesses only a rather poorly developed mind to make up.

All this new and important scientific description is not merely presented with the most lucid simplicity, but enlivened with some extremely entertaining embellishments. Poor Lorenz being forced to spend hours crouched on his knees or crawling on hands and feet, and quacking loudly at frequent intervals, if he was to fulfil his role as "imprinted" parent of a brood of ducklings; his assistant suddenly realizing he was talking goose instead of duck to the same ducklings, and cutting short his goose-talk with "No, I mean *quah, quah, quah*"; Lorenz's old father walking back to the house from his outdoor siesta, indignantly holding up his trousers because Lorenz's tame cockatoo had bitten all the buttons off all his clothes—coat buttons, waistcoat buttons, braces buttons and fly buttons—and laid them out in order on the ground; Lorenz calling down the same cockatoo from high up in the air by emitting repeated cockatoo-screams (visitors to the parrot house at the Zoo will remember what that means!) on a crowded railway platform—these and various other incidents that he records I shall long chuckle over.

But I do not wish to stand between Lorenz and his readers. I will conclude by expressing my fullest agreement with him when he repudiates the unimaginative and blinkered outlook of those who think that it is "scientific" to pretend that something rich and complex is merely its jejune and simple elements, and in particular that the brains of higher organisms, such as birds, those complex body-minds with their elaborate emotional behaviour, are "really" nothing but reflex machines, like a bit of special cord magnified and supplied with special sense-organs; and equally so when he repudiates the uncritical and often wishful thinking of the sentimental anthropomorphizers, who not merely refuse to take the trouble to understand the radically different nature of animals' minds and behaviour from our own, but in fact are satisfying some repressed urge of their own unconscious by projecting human attributes into bird and beast.

As he rightly says, the truth is more extraordinary and more interesting than any such futile imaginings. He might have added that the truth is also necessary. Only if we know and face the truth about the world, whether the world of physics and chemistry, or of geology and biology, or of mind and behaviour, shall we be able to see what is our own true place in that world. Only as we discover and assimilate the truth about nature shall we be able to undertake the apparently contradictory but essential task of re-establishing our unity with nature while at the same time maintaining our transcendence over nature. The work of men like Lorenz is a very real contribution to our understanding of our relations with that important part of nature constituted by the higher animals.

PREFACE

There was never a king like Solomon
Not since the world began,
Yet Solomon talked to a butterfly
As a man would talk to a man.

RUDYARD KIPLING

AS Holy Scripture tells us, the wise King Solomon, the son of David, "spake also of beasts, and of fowl, and of creeping things, and of fishes" (I Kings IV. 33). A slight misreading of this text, which very probably is the oldest record of a biological lecture, has given rise to the charming legend that the King was able to talk the language of animals, which was hidden from all other men. Although this venerable tale that he spake to the animals and not of them certainly originated from a misunderstanding, I feel inclined to accept it as a truth; I am quite ready to believe that Solomon really could do so, even without the help of the magic ring which is attributed to him by the legend in question, and I have very good reason for crediting it; I can do it myself, and without the aid of magic, black or otherwise. I do not think it is very sporting to use magic rings in dealing with animals. Without supernatural assistance, our fellow creatures can tell us the most beautiful stories, and that means *true* stories, because the truth about nature is always far more beautiful even than what our great poets sing of it, and they are the only real magicians that exist.

I am not joking by any means. In so far as the "signal code" of a species of social animal can be called a language at all, it can be understood by a man who has got to know its "vocabulary", a subject to which a whole chapter in this book is devoted. Of course lower and non-social animals do not have anything that could, even in a very wide sense, be compared with a language, for the very simple reason that they do not have anything to say. For the same reason, it is impossible to say anything to them; it would, indeed, be exceedingly difficult to say anything that would interest some of the lower "creeping things". But by knowing the "vocabulary" of some highly social species of beast or bird it is often possible to attain to an astonishing intimacy and mutual understanding. In the day's work of a scientist investigating animal behaviour this becomes a matter of course and ceases to be a source of wonder, but I still retain the clear-cut memory of a very funny episode, which, with all the suddenness of philosophical realization, brought to my full consciousness what an astounding and unique thing the close social relation between a human and a wild animal really is.

Before I begin, I must first of all describe the setting which forms the background for most of this book. The beautiful country flanking the Danube on either side in the district of Altenberg is a real "naturalist's paradise". Protected against civilization and agriculture by the yearly inundations of Mother Danube, dense willow forests, impenetrable scrub, reed-grown marshes and drowsy backwaters stretch over many square miles; an island of utter wildness in the middle of Lower Austria; an oasis of virgin nature, in which red and roe deer, herons and cormorants have survived the vicissitudes even of the last terrible war. Here, as in Wordsworth's beloved Lakeland,

> The duck dabbles mid the rustling sedge
> And feeding pike starts from the water's edge
> And heron, as resounds the trodden shore,
> Shoots upward, darting his long neck before.

The virgin wildness of this stretch of country is something rarely found in the very heart of old Europe. There is a strange contrast between the character of the landscape and its geographical situation, and, to the naturalist's eye, this contrast is emphasized by the presence of a number of American plants and animals which have been introduced. The American golden rod (*Solidago virgoaurea*) dominates the landscape above water, as does *Elodea canadensis* below the surface: American sun perch (*Eupomotis gibbosus*) and catfish (*Amiurus nebulosus*) are common in some backwaters; and something heavy and ponderous in the figure of our stags betrays, to the initiated, that Francis Joseph I, in the heyday of his hunting life, introduced a few hundred head of wapiti to Austria. Muskrats are abundant, having made their way down from Bohemia, where they were first released in Europe, and the loud splash of their tails, when they smack the surface of the water as a warning signal, mingles with the sweet notes of the European oriole.

To all this, you must add the picture of Mother Danube, who is little sister to the Mississippi, and imagine the river itself with its broad, shallow, winding bed, its narrow navigable channel that changes its course continuously, unlike all other European rivers, and its mighty expanse of turbulent waters that alter their colours with the season, from turbid greyish-yellow in spring and summer to clear blue-green in late autumn and winter. The "Blue Danube", made famous by our popular songs, exists only in the cold season.

Now imagine this queerly mixed strip of river landscape as being bordered by vine-covered hills, brothers to those flanking the Rhine, from whose crests the two early mediæval castles of Greifenstein and Kreuzenstein look down with serious mien over the vast expanse of wild forest and water. Then you have before you the landscape which is the setting of this story-book, the landscape which I consider the most beautiful on earth, as every man should consider his own home country.

One hot day in early summer, when my friend and assistant Dr Seitz and I were working on our greylag goose

film, a very queer procession slowly made its way through this beautiful landscape, a procession as wildly mixed as the landscape itself. First came a big red dog, looking like an Alaskan husky, but actually a cross between an Alsatian and a Chow, then two men in bathing-trunks carrying a canoe, then ten half-grown greylag goslings, walking with all the dignity characteristic of their kind, then a long row of thirteen tiny cheeping mallard ducklings, scurrying in pursuit, forever afraid of being lost and anxiously striving to keep up with the larger animals. At the end of the procession marched a queer piebald ugly duckling, looking like nothing on earth, but in reality a hybrid of ruddy sheldrake and Egyptian goose. But for the bathing-trunks and the moving-picture camera slung across the shoulders of one of the men, you might have thought you were watching a scene out of the Garden of Eden.

We progressed very slowly, as our pace was set by the weakest among our little mallards, and it took us some considerable time to get to our destination, a particularly picturesque backwater, framed by blossoming snowballs and chosen by Seitz to "shoot" certain scenes of our greylag film. When we arrived, we at once got down to

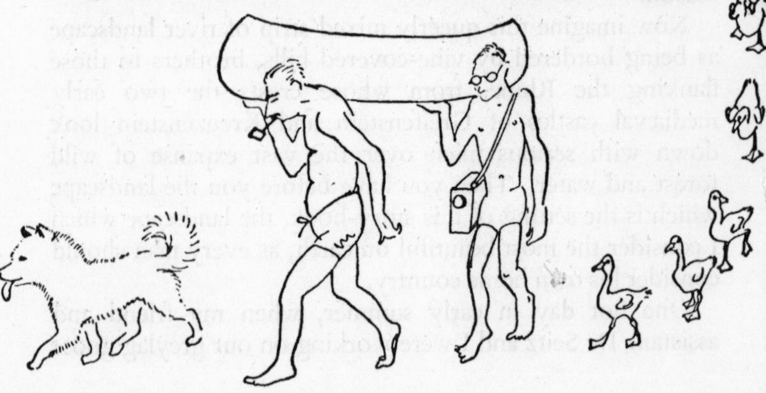

business. The title of the film says "Scientific direction: Dr Konrad Lorenz. Camera: Dr Alfred Seitz". Therefore I at once proceeded to direct scientifically, this for the moment consisting in lying down on the soft grass bordering the water and sunning myself. The green water-frogs were croaking in the lazy way they have on summer days, big dragon-flies came whirling past and a black-cap warbled its sweetly jubilant song in a bush not three yards from where I lay. Farther off, I could hear Alfred winding up his camera and grumbling at the little mallards who forever kept swimming into the picture, while for the moment he did not want anything in it but greylags. In the higher centres of my brain, I was still aware that I ought to get up and help my friend by luring away the mallards and the Ruddy Egyptian, but although the spirit was willing the flesh was weak, for exactly the same reason as was that of the disciples in Gethsemane: I was falling asleep. Then suddenly, through the drowsy dimness of my senses, I heard Alfred say, in an irritated tone: "Rangangangang, rangangangang—oh, sorry, I mean—quahg, gegegegeg, Quahg, gegegegeg!" I woke laughing: he had wanted to call away the mallards and had, by mistake, addressed them in greylag language.

It was at that very moment that the thought of writing a book first crossed my mind. There was nobody to appreciate the joke, Alfred being far too preoccupied with his work. I wanted to tell it to somebody and so it occurred to me to tell it to everybody.

And why not? Why should not the comparative ethologist who makes it his business to know animals more thoroughly than anybody else, tell stories about their private lives? Every scientist should, after all, regard it as his duty to tell the public, in a generally intelligible way, about what he is doing.

There are already many books about animals, both good and bad, true and false, so one more book of true stories cannot do much harm. I am not contending, though, that a good book must unconditionally be a true one. The mental development of my own early childhood was, without any doubt, influenced in a most beneficial way by two books of animal stories which cannot, even in a very loose sense, be regarded as true. Neither Selma Lagerlof's *Nils Holgersson* nor Rudyard Kipling's *Jungle Books* contains anything like scientific truth about animals. But poets such as the authors of these books may well avail themselves of poetic licence to present the animal in a way far divergent from scientific truth. They may daringly let the animal speak like a human being, they may even ascribe human motives to its actions, and yet succeed in retaining the general style of the wild creature. Surprisingly enough, they convey a true impression of what a wild animal is like, although they are telling fairy tales. In reading those books, one feels that if an experienced old wild goose or a wise black panther could talk, it would say exactly the things which Selma Lagerlof's Akka or Rudyard Kipling's Bagheera say.

The creative writer, in depicting an animal's behaviour, is under no greater obligation to keep within the bounds of exact truth than is the painter or the sculptor in shaping an animal's likeness. But all three artists must regard it as their most sacred duty to be properly instructed regarding those particulars in which they deviate from the actual facts. They must indeed be even better informed on these details than on others which they render in a manner true to nature. There is no greater sin against the spirit of true art, no more contemptible dilettantism than to use artistic licence as a specious cover for ignorance of fact.

I am a scientist and not a poet and I shall not aspire, in this little book, to improve on nature by taking any artistic liberties. Any such attempt would certainly have the opposite effect, and my only chance of writing something not entirely devoid of charm lies in strict adherence to scientific fact. Thus, by modestly keeping to the methods of my own craft, I may hope to convey, to my kindly reader, at least a slight inkling of the infinite beauty of our fellow creatures and their life.

KONRAD Z. LORENZ

Altenberg, January 1950

1. ANIMALS AS A NUISANCE

Split open the kegs of salted sprats,
Made nests inside men's Sunday hats,
And even spoiled the women's chats,
By drowning their speaking
With shrieking and squeaking
In fifty different sharps and flats.

ROBERT BROWNING

WHY should I tell first of the darker side of life with animals? Because the degree of one's willingness to bear with this darker side is the measure of one's love for animals. I owe undying gratitude to my patient parents, who only shook their heads or sighed resignedly when, as a schoolboy or young student, I once again brought home a new and probably yet more destructive pet. And what has my wife put up with, in the course of the years? For who else would dare ask his wife to allow a tame rat to run free around the house, gnawing neat little circular pieces out of the sheets to furnish her nests, which she built in even more awkward places than men's Sunday hats?

Or what other wife would tolerate a cockatoo who bit off all the buttons from the washing hung up to dry in the garden, or allow a greylag goose to spend the night in the bedroom and leave in the morning by the window? (Greylag geese cannot be house-trained.) And what would she say when she found out that the nice little blue spots with which song-birds, after a repast of elderberries, decorate all the furniture and curtains, just will not come out in the wash? What would she say, if . . . I could go on asking for twenty pages!

Is all this absolutely necessary? Yes, quite definitely yes! Of course one can keep animals in cages fit for the drawing-room, but one can only get to know the higher and mentally active animals by letting them move about freely. How sad and mentally stunted is a caged monkey or parrot, and how incredibly alert, amusing and interesting is the same animal in complete freedom. Though one must be prepared for the damage and annoyance which are the price one has to pay for such house-mates, one obtains a mentally healthy subject for one's observations and experiments. This is the reason why the keeping of higher animals in a state of unrestricted freedom has always been my speciality.

In Altenberg the wire of the cage always played a paradoxical role: it had to prevent the animals entering the house or front garden. They were also strictly forbidden to go within the wire netting that fenced in our flower-beds; but forbidden things have a magnetic attraction for intelligent animals, as for little children. Besides, the delightfully affectionate greylag geese long for human society. So it was always happening that, before we had noticed it, twenty or thirty geese were grazing on the flower-beds, or, worse still, with loud honking cries of greeting, had invaded the closed-in veranda. Now it is uncommonly difficult to repel a bird which can fly, and has no fear of man. The loudest shouts, the wildest waving of arms have no effect whatever. Our only really effective scarecrow was an enormous scarlet garden umbrella. Like a knight with lance at rest, my wife would tuck the folded umbrella under her arm and spring at the geese who were again grazing on her freshly planted beds; she would let out a frantic war-cry and unfold the umbrella with a sudden jerk; that was too much even for our geese, who, with a thundering of wings, took to the air.

Unfortunately, my father largely undid all my wife's efforts in goose education. The old gentleman was very fond of the geese, and he particularly liked the ganders for their courageous chivalry; so nothing could prevent him from inviting them, each day, to tea in his study adjoining the glass veranda. As, at this time, his sight was already failing, he only noticed the material result of such a visit when he trod right into it. One day, as I went into the garden, towards the evening, I found, to my astonishment, that nearly all the grey geese were missing. Fearing the worst, I ran to my father's study, and what did I see? On the beautiful Persian carpet stood twenty-four geese, crowded round the old gentleman who was drinking tea at his desk, quietly reading the newspaper and holding out to the geese one piece of bread after another. The birds were somewhat nervous in their unaccustomed surroundings and this, unfortunately, had an adverse effect on their intestinal movements, for, like all animals that have to digest much grass, the goose has a cæcum or blind appendage of the large intestine in which vegetable fibre is made assimilable for the body by the action of cellulose-splitting bacteria. As a rule, to about six or seven normal evacuations of the intestine there occurs one of the cæcum, and this has a peculiar pungent smell and a very bright dark green colour. If a goose is nervous, one cæcum evacuation follows after another. Since this goose tea-party more than eleven years have elapsed; the dark green stains on the carpet have meanwhile become pale yellow.

So the animals lived in complete freedom and yet in great familiarity with our house. They always strove towards us instead of away from us. In other households, people might call: "The bird has escaped from its cage; quick, shut the window!" But with us the cry was: "For goodness' sake, shut the window; the cockatoo (raven, monkey, etc.) is trying to get in!" The most paradoxical use of the "inverse cage principle" was invented by my wife when our eldest son was very small. At that time we kept several large and potentially dangerous animals—some ravens, two greater yellow-crested cockatoos, two Mongoz Lemurs, and two capuchin monkeys, none of which could safely be left alone with the child. So my wife improvised, in the garden, a large cage, and inside it she put . . . the pram.

In the higher animals the ability and inclination to do damage are, unfortunately, in direct proportion to the degree of their intelligence. For this reason, it is impossible to leave certain animals, particularly monkeys, permanently loose and without supervision. With lemurs, however, this is possible, since they lack that searching curiosity which all true monkeys display in respect of household implements. True monkeys, on the other hand, even the genealogically lower-standing New World monkeys (*Platyrrhinae*), have an insatiable curiosity for every object that is new to them, and they proceed to experiment with it. Interesting though that may be from the standpoint of the animal psychologist, for the household it soon becomes a financially unbearable state of affairs. I can illustrate this with an example.

As a young student, I kept, in my parents' flat in Vienna, a magnificent specimen of a female capuchin monkey named

Gloria. She occupied a large, roomy cage in my study. When I was at home and able to look after her, she was allowed to run freely about the room. When I went out, I shut her in the cage, where she became exceedingly bored and exerted all her talents to escape as quickly as possible. One evening, when I returned home after a longer absence and turned the knob of the light-switch, all remained dark as before. But Gloria's giggle, issuing not from the cage but from the curtain rod, left no doubt as to the cause and origin of the light defect. When I returned with a lighted candle, I encountered the following scene: Gloria had removed the heavy bronze bedside lamp from its stand, dragged it straight across the room (unhappily without pulling the plug out of the wall), heaved it up on the highest of my aquaria, and, as with a battering ram, bashed in the glass lid so that the lamp sank in the water. Hence the short circuit! Next, or perhaps earlier, Gloria had unlocked my bookcase —an amazing achievement considering the minute size of the key—removed volumes 2 and 4 of Strumpel's textbook of medicine and carried them to the aquarium stand, where she tore them to shreds and stuffed them into the tank. On the floor lay the empty book-covers, but not one piece of paper. In the tank sat sad sea-anemones, their tentacles full of paper. . . .

The interesting part of these proceedings was the strict attention to detail with which the whole business had been performed: Gloria must have dedicated considerable time to her experiments: physically alone, this accomplishment was, for such a small animal, worthy of recognition: only rather expensive.

But what are the positive values that redeem all this endless annoyance and expense? We have already mentioned that it is necessary, for certain observations, to have an animal that is not a prisoner. Apart from this, the animal that could escape and yet remains with me affords me undefinable pleasure, especially when it is affection for myself that has prompted it to stay.

On one occasion, while walking near the banks of the Danube, I heard the sonorous call of a raven, and when, in response to my answering cry, the great bird, far up in the sky, folded his wings, came whizzing down at breathless speed, and with a rush of air checked his fall on outstretched pinions, to land on my shoulder with weightless ease, I felt compensated for all the torn-up books and all the plundered duck nests that this raven of mine had on his conscience. The magic of such an experience is not blunted by repetition; the wonder of it remains, even when it is an everyday occurrence and Odin's bird has become, for me, as natural a pet as, to anyone else, a dog or cat. Real friendship with wild animals is to me so much a matter of course that it takes special situations to make me realize its uniqueness. One misty spring morning I went down to the Danube. The river was still shrunk to its winter proportions, and migrating golden-eyes, mergansers, smews and here and there a flock of bean- or white-fronted geese came flying along its dark and narrowed surface. Among these migrants, quite as if they belonged to them, a flock of greylag geese winged its

way. I could see that the goose flying second on the left of the triangular phalanx had lost a primary. And in this moment there flashed across my inward eye vivid reminiscences of this goose with its missing primary and of all that had happened when it was broken. For, of course, these were my greylag geese; there are indeed no others on the Danube, even at migration time. The second bird on the left wing of the triangle was the gander Martin. He had just got engaged to my pet goose Martina and was therefore christened after her (formerly he was just a number, because only the geese reared by myself received names, while those that were brought up by their parents were numbered). In greylag geese, the young bridegroom follows literally in the footsteps of his bride, but Martina wandered free and fearless through all the rooms of our house, without stopping to ask the advice of her bridegroom, who had grown up in the garden; so he was forced to venture into realms unknown to him. If one considers that a greylag goose, naturally a bird of open country, must overcome strong instinctive aversions in order to venture even between bushes or under trees, one is forced to regard Martin as a little hero as, with upstretched neck, he followed his bride through the front door into the hall and then upstairs into the bedroom. I see him now standing in the room, his feathers flattened against his body with fear, shivering with tension, but proudly erect and challenging the great unknown by loud hisses. Then suddenly the door behind him shut with a bang. To remain steadfast now was too much to ask even of a greylag goose hero. He spread his wings and flew, straight as a die, into the chandelier. The latter lost a few appendages, but Martin lost a primary.

So that is how I know about the missing feather of the goose flying second in the left wing of the triangle; but I know, too, something that is truly comforting: when I come home from my walk, these grey geese, now flying in company with wild migrants, will be standing on the steps in front of the veranda and they will come to greet me, their necks outstretched in that gesture which, in geese, means the same as tail-wagging in a dog. And, as my eyes follow the geese, which, flying low over the water, disappear round the next bend of the river, I am all at once gripped by amazement as, with that wonderment which is the birth-act of philosophy, I suddenly start to query the familiar. We have all experienced that deeply moving sensation in which the most everyday things suddenly stare us in the face with altered mien as though we were seeing them for the first time. Wordsworth became conscious of this one day while contemplating the Lesser Celandine:

> I have seen thee, high and low,
> Thirty years or more, and yet
> 'Twas a face I did not know;
> Thou hast now, go where I may,
> Fifty greetings in a day.

As I watched the geese, it appeared to me as little short of a miracle that a hard, matter-of-fact scientist should have been able to establish a real friendship with wild, free-living animals, and the realization of this fact made me strangely happy. It made me feel as though man's expulsion from the Garden of Eden had thereby lost some of its bitterness.

To-day the ravens are gone, the greylag geese were scattered by the war. Of all my free-flying birds, only the jackdaws remain; they were the first of all the birds that I installed in Altenberg. These perennial retainers still circle round the high gables, and their shrill cries, whose meaning I understand in every detail, still echo through the shafts of the central heating into my study. And every year they stop up the chimneys with their nests and infuriate the neighbours by eating their cherries.

Can you understand that it is not only scientific results that are the recompense for all this trouble and annoyance, but more, much, much more?

meant adjustments on the part of the owner may cause much damage. It is, of course, possible to set up a "pretty" aquarium with artificial foundations and carefully distributed plants; a filter would prevent any mud formation, and artificial aeration permit the keeping of many more fish than would otherwise be possible. In this case the plants are merely ornamental: the animals do not require them, since they derive from the artificial aeration enough oxygen for their maintenance. It is purely a matter of taste, but I personally think of an aquarium as of a living community that regulates its own equilibrium. The other kind is a "cage", an artificially cleaned container which is not an end in itself, but purely a means of keeping certain animals.

It is a real art to determine in advance the type of animal and plant community which one wants to develop in an aquarium, and to do this requires much experience and biological tact in choosing the right materials for the bottom, the situation of the tank, the heat and light conditions and finally the plant and animal inmates themselves. A past

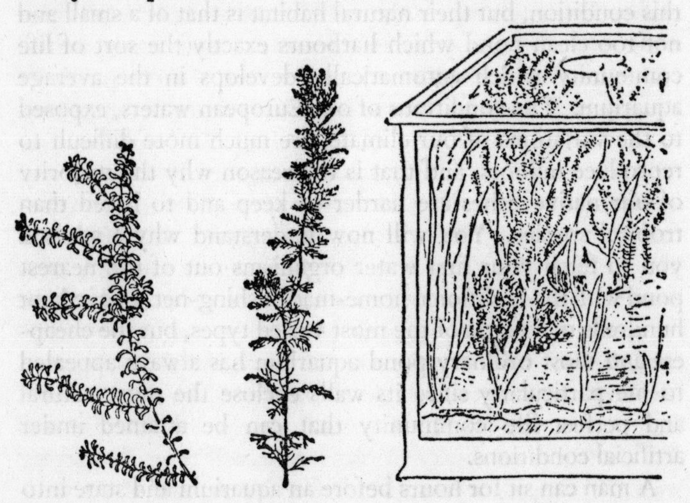

master of this art was my tragically deceased friend Bernhard Hellman, who was able to copy, at will, any given type of pond or lake, brook or river. One of his masterpieces was a large aquarium which was a perfect model of an Alpine lake. The tank was very deep and cool and was placed not too near the light, the vegetation in the crystal-clear water consisted of glassily transparent, pale green pond weed (*Potamogeton*), the stony bottom was covered with dark green Fontinalis and decorative stonewort (*Chara*). Of the non-microscopic animals the only representatives were some minute trout and minnows, a few freshwater shrimps and a little crayfish. Thus, the animal inhabitants were so few that they hardly required feeding, since they were able to subsist on the natural microfauna of the aquarium.

If one wishes to breed some of the more delicate water animals, it is essential, in the construction of an aquarium, to reproduce the whole of the natural habitat with its entire community of living macro- and micro-organisms. Even the commonest of tropical aquarium fishes are dependent on this condition, but their natural habitat is that of a small and not too clean pond which harbours exactly the sort of life community which automatically develops in the average aquarium. The conditions of our European waters, exposed to the variations of our climate, are much more difficult to reproduce indoors, and that is the reason why the majority of our native fishes are harder to keep and to breed than tropical species. You will now understand why I advised you to fetch your first water organisms out of the nearest pond with the traditional home-made fishing-net. I have kept hundreds of aquaria of the most varied types, but the cheapest and most ordinary pond aquarium has always appealed to me particularly since its walls enclose the most natural and perfect life community that can be attained under artificial conditions.

A man can sit for hours before an aquarium and stare into

it as into the flames of an open fire or the rushing waters of a torrent. All conscious thought is happily lost in this state of apparent vacancy, and yet, in these hours of idleness, one learns essential truths about the macrocosm and the microcosm. If I cast into one side of the balance all that I have learned from the books of the library and into the other everything that I have gleaned from the "books in the running brooks", how surely would the latter turn the scales.

3. ROBBERY IN THE AQUARIUM

How cheerfully he seems to grin,
How neatly spreads his claws,
And welcomes little fishes in
With gently smiling jaws!

LEWIS CARROLL, *Alice in Wonderland*

THERE are some terrible robbers in the pond world, and in our aquarium we may witness all the cruelties of an embittered struggle for existence enacted before our very eyes. If you have introduced to your aquarium a mixed catch, you will soon be able to see an example of these conflicts, for amongst the new arrivals there will probably be a larva of the water-beetle Dytiscus. Considering their relative size, the voracity and cunning with which these animals destroy their prey eclipse the methods of even such notorious robbers as tigers, lions, wolves or killer whales. These are all as lambs compared with the Dytiscus larva.

It is a slim, streamlined insect, rather more than two inches long. Its six legs are equipped with stout fringes of bristles which form broad oar-like blades that propel the animal with quick and sure movements through the water. The wide, flat head bears an enormous, pincer-shaped pair of jaws which are hollow and serve not only as syringes for injecting poison, but also as orifices of ingestion. The animal lies in ambush on some water-plant; suddenly it shoots at lightning speed towards its prey, darts underneath it, then quickly jerks up its head and grabs the victim in its jaws. "Prey", for these creatures, is all that moves or that smells of "animal" in any way. It has often happened to me that, while standing quietly in the water of a pond, I have been "eaten" by a Dytiscus larva. Even for man, an injection of the poisonous digestive juice of this insect is extremely painful.

These beetle larvæ are among the few animals which digest "out of doors". The glandular secretion that they inject, through their hollow forceps, into their prey, dissolves the entire inside of the latter into a liquid soup which is then sucked in through the same channel by the attacker. Even large victims, such as fat tadpoles or dragon-fly larvæ, which have been bitten by a Dytiscus larva, stiffen after a few defensive movements, and their inside, which, as in most water animals, is more or less transparent, becomes opaque as though fixed by formalin. The animal swells up first, then gradually shrinks to a limp bundle of skin which hangs from the deadly jaws, and is finally allowed to drop. In the confined spaces of an aquarium, a few large Dytiscus larvæ will, within a few days, eat all living things over about a quarter of an inch long. What happens then? They will eat each other, if they have not already done so; this depends less on who is bigger and stronger than upon who succeeds in seizing the other first. I have often seen two nearly equal-sized Dytiscus larvæ each seize the other simultaneously and both die a quick death by inner dissolution. There are very few animals which, even when threatened with starvation, will attack an equal-sized animal of their own species with the intention of devouring it. I only know this to be definitely true of rats and a few related rodents; that wolves do the same thing I am much inclined to doubt, on the strength of some observations of which I shall speak later. But Dytiscus larvæ devour animals of their own breed and size, even when other nourishment is at hand, and that is done, as far as I know, by no other animal.

A somewhat less brutal but more elegant beast of prey is the larva of the great dragon-fly Aeschna. The mature insect is a true king of the air, a veritable falcon among insects, for it catches its prey when on the wing. If you shake your pond catch into a wash-basin, in order to remove the worst miscreants, you will possibly find, besides Dytiscus larvæ, some other streamlined insects whose remarkable method of locomotion at once attracts the attention. These slender torpedoes, which are usually marked with a decorative pattern of yellow and green, shoot forward in rapid jerks, their legs pressed close to their sides. It is at first something of an enigma how they move at all. But if you observe them separately, in a shallow dish, you will see that these larvæ are jet-propelled. From the tip of their abdomen there squirts forth a powerful little column of water which drives the animal speedily forward. The end portion of their intestine forms a hollow bladder which is richly lined with tracheal gills and serves at the same time the purposes of respiration and of locomotion.

Aeschna larvæ do not hunt swimming but lie in ambush: when an object of prey comes within eye range they fix it with their gaze, turn their head and body very slowly in its direction and follow all its movements attentively. This marking down of the prey can only be observed in a very few other non-vertebrate animals. In contrast to the larvæ of Dytiscus, those of Aeschna can see even very slow movements, such as the crawling of snails which therefore very often fall a prey to them. Slowly, very slowly, step by step, the Aeschna larva stalks its prey: it is still an inch or two away when suddenly—what was that?—the victim is struggling between the cruel jaws. Without taking a slow-motion picture of this procedure, one could only see that something tongue-like flew out from the head of the larva to its prey and drew the latter instantly within reach of the attacker's jaws. Anyone who had ever seen a chameleon eating would at once be reminded of the flicking back and forth of its sticky tongue. The "boomerang" of the Aeschna is, however, no tongue but the metamorphosed "underlip", which consists of two movable joints with a pincer at their end.

The optical fixation of its prey alone makes the dragon-fly larva appear strangely "intelligent", and this impression will be strengthened should some other peculiarities of its behaviour be observed. In contrast to the Dytiscus larva, which will snap blindly at anything, the dragon-fly larva leaves animals above a certain size severely alone, even if it has been starving for weeks. I have kept Aeschna larvæ for months in a basin with fish, and have never seen them attack or damage one larger than themselves. It is a remarkable fact that the larvæ will never grab at a prey which has been caught by a member of their own species and which is now moving slowly backwards and forwards between the masticating jaws; on the other hand they will at once take a piece of fresh meat moved in a like manner on the end of a glass feeding rod in front of their eyes. In my large American sun-perch aquarium I always had a few Aeschna larvæ growing up: their development takes long—more than a year. Then, on a beautiful summer's day, comes the great moment; the larva climbs slowly up the stem of a plant and out of the water. There it sits for a long time and then, as in every moulting process, the outer skin on the back part of the thoracic segments bursts open and the beautiful perfect insect unwinds itself slowly from the larval skin. After this, several hours expire before the wings have reached their full size and consistency, and this is attained by a wonderful process whereby a rapidly solidifying liquid is pumped, under high pressure, into the fine branches of the wing veins. Then you open the window wide and wish your aquarium guest good luck and *bon voyage* in its insect life.

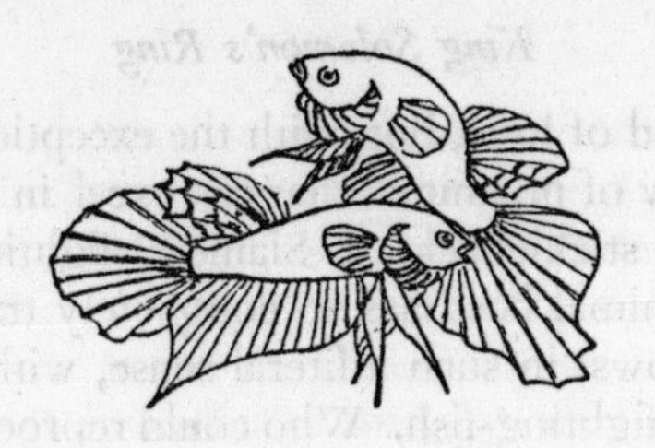

4. POOR FISH

Weed in the wave, gleam in the mud—
The dark fire leaps along his blood;
Dateless and deathless, blind and still,
The intricate impulse works its will.

RUPERT BROOKE, *The Fish*

STRANGE what blind faith is placed in proverbs, even when what they say is false or misleading. The fox is not more cunning than other beasts of prey and is much more stupid than wolf or dog, the dove is certainly not peaceful, and of the fish, rumour spreads only untruth: it is neither so cold-blooded as one says of dull people, nor is the "fish in water" nearly so happily situated as the converse saying would imply. In reality there is no other group of animals that, even in nature, is so plagued with infectious diseases as the fish. I have never yet known a newly caught bird, reptile or mammal bring an infectious disease into my animal population; but very newly acquired fish must, as a routine measure, go into the quarantine aquarium, otherwise you may bet a hundred to one that within a very short time the dreaded minute white spots, the sign of infection with the parasite *Ichthyophtirius multifiliis*, will appear on the fins of the previously installed aquarium dwellers.

And regarding the alleged cold-bloodedness of fishes; I am familiar with many animals and with their behaviour in the most intimate situations of their life, in the wild ecstasies of the fight and of love, but, with the exception of the wild canary, I know of no animal that can excel in hot-bloodedness a male stickleback, a Siamese fighting-fish or a cichlid. No animal becomes so completely transformed by love, none glows, in such a literal sense, with passion as a stickleback or fighting-fish. Who could reproduce in words, what artist in colour, that glowing red that makes the sides of the male stickleback glassy and transparent, the iridescent blue-green of its back whose colour and brilliance can only be compared with the illuminating power of neon lighting, or finally, the brilliant emerald-green of its eyes? According to the rules of artistic taste, these colours should clash horribly, and yet what a symphony they produce, composed by the hand of nature.

In the fighting-fish, this marvel of colour is not continually present. For the little brown-grey fish that lies with folded fins in one corner of the aquarium reveals nothing of it for the moment. It is only when another fish, equally inconspicuous at first, approaches him and each sights the other, that they begin gradually to light up in all their incandescent glory. The glow pervades their bodies almost as quickly as the wire of an electric heater grows red. The fins unfold themselves like ornamental fans, so suddenly that one almost expects to hear the sound of an umbrella being opened quickly. And now follows a dance of burning passion, a dance which is not play but real earnest, a dance of life or death, of be-all or end-all. To begin with, strangely enough, it is uncertain whether it will lead to love overtures and mating, or whether it will develop, by an equally flowing transition, into a bloody battle. Fighting-fish recognize the sex of a member of their own species not simply by seeing it but by watching the way in which it responds to the severely ritualized, inherited, instinctive movements of the dancer.

The meeting of two previously unacquainted fighting-fish begins with a mutual "showing-off", a swaggering act of self-display in which every luminous colour-spot and every iridescent ray of the wonderful fins is brought into maximum play. Before the glorious male, the modestly garbed female lowers the flag—by folding her fins—and, if she is unwilling to mate, flees immediately. Should she be willing to mate, she approaches the male with shy, insinuating movements, that is to say, in an attitude directly opposed to that of the swaggering male. And now begins a love ceremonial which, if it cannot compare in grandeur with the male war-dance, can emulate it in grace of movement.

When two males meet face to face, veritable orgies of mutual self-glorification take place. There is a striking similarity between the war-dance of these fish and the corresponding ceremonial dances of Javanese and other Indonesian peoples. In both man and fish the minutest detail of every movement is laid down by immutable and ancient laws, the slightest gesture has its own deeply symbolic meaning. There is a close resemblance between man and fish in the style and exotic grace of their movements of restrained passion.

The beautifully refined form of the movements betrays the fact that they have a long historical development behind them and that they owe their elaborateness to an ancient ritual. It is, however, not so obvious that though in man this ritual is a ceremony which has been handed down from generation to generation by a thousand-year-old tradition, in the fish it represents the result of an evolutional development of innate instinctive activities, at least a hundred times older. Genealogical research into the origin of such ritual expression and the comparison of such ceremonies in related species are exceedingly illuminating. We know more of the evolutionary history of these movements than of all other instincts.

After this digression, let us return to the war-dance of the male fighting-fish. This has exactly the same meaning as the duel of words of the Homeric heroes, or of our Alpine farmers, which, even to-day, often precedes the traditional Sunday brawl in the village inn. The idea is to intimidate one's opponent and at the same time to stimulate oneself to a state of fearlessness. In the fish, the long duration of these preliminaries, their ritual character and above all their great show of colour finery and fin development, which at first only serve to subdue the opponent, mask, for the uninitiated, the seriousness of the situation. On account of their beauty, the fighters appear less malevolent than they really are, and one is just as loath to ascribe to them embittered courage and contempt of death as one is to associate head-hunting with the almost effeminately beautiful Indonesian warriors. Nevertheless both are capable of fighting to the death. The battles of the fighting-fish often end in the death of one of the adversaries. When they are stimulated to the point of inflicting the first sword-thrust, it is only a matter of minutes till wide slits are gaping in their fins, which in a few more minutes are reduced to tatters. The method of attack of a fighting-fish, as of nearly all fish that fight, is literally the sword-thrust and not the bite. The fish opens its jaws so wide that all its teeth are directed forwards, and in this attitude it rams them, with all the force of its muscular body, into the side of its adversary. The ramming of a fighting-fish is so strong and hard that its impact is clearly audible if, in the confusion of the fight, one of the antagonists happens to hit the glass side of the tank. The self-display dance can last for hours but, should it develop into action, it is often only a matter of minutes before one of the combatants lies mortally wounded on the bottom.

The fights of our European sticklebacks are very different

from those of the Siamese fighting-fish. In contrast to the latter, the stickleback, at mating time, glows not only when it sees an opponent or a female, but does so as long as it is in the vicinity of its nest, in its own chosen territory. The basic principle of his fighting is, "my home is my castle". Take his nest from a stickleback or remove him from the tank where he built it and put him with another male, and he will not dream of fighting but, on the contrary, will make himself small and ugly. It would be impossible to use sticklebacks for exhibition battles as the Siamese have done, for hundreds of years, with fighting-fish. It is only when he has founded his home that the stickleback becomes physically capable of reaching a state of full sexual excitement; therefore, a real stickleback fight can be seen only when two males are kept together in a large tank where they are both building their nests. The fighting inclinations of a stickleback, at any given moment, are in direct proportion to his proximity to his nest. At the nest itself he is a raging fury and with a fine contempt of death will recklessly ram the strongest opponent, or even the human hand. The farther he strays from his headquarters in the course of his swimming, the more his courage wanes. When two sticklebacks meet in battle it is possible to predict with a high degree of certainty how the fight will end: the one which is farther from his nest will lose the match. In the immediate neighbourhood of his nest, even the smallest male will defeat the largest one, and the relative fighting potential of the individual is shown by the size of the territory which he can keep clear of rivals. The vanquished fish invariably flees homeward and the victor, carried away by his successes, chases the other furiously, far into its domain. The farther the victor goes from home, the more his courage ebbs, while that of the vanquished rises in proportion. Arrived in the precincts of his nest, the fugitive gains new strength, turns right about and dashes with gathering fury at his pursuer. A new battle begins, which ends with absolute certainty in the defeat of the former victor, and off goes the chase again in the opposite direction. The pursuit is repeated a few times in alternating directions, swinging to and fro like a pendulum which at last reaches a state of equilibrium at a certain point. The line at which the fighting potentials of the individuals are thus equally balanced marks the border of their territories. This same principle is of great importance in the biology of many animals, particularly that of birds. Every bird lover has seen two male redstarts chasing each other in exactly the same manner.

Once on this borderline, both sticklebacks hesitate to attack. Taking on a peculiar threatening attitude, they incessantly stand on their heads and, like Father William, they do it again and again. At the same time they turn broadside on towards each other, and each erects threateningly the ventral spine on the side nearer his opponent. All the while they seem to be "pecking" at the bottom for food. In reality, however, they are executing a ritualized version of the activity normally used in nest-digging. If an animal finds the outlet for some instinctive action blocked by a conflicting drive, it often finds relief by discharging an entirely different instinctive movement. In this case, the stickleback, not quite daring to attack, finds an outlet in nest-digging. This type of phenomenon, which is of great theoretical interest both from the physiological and psychological point of view, is termed in comparative ethology a "displacement activity".

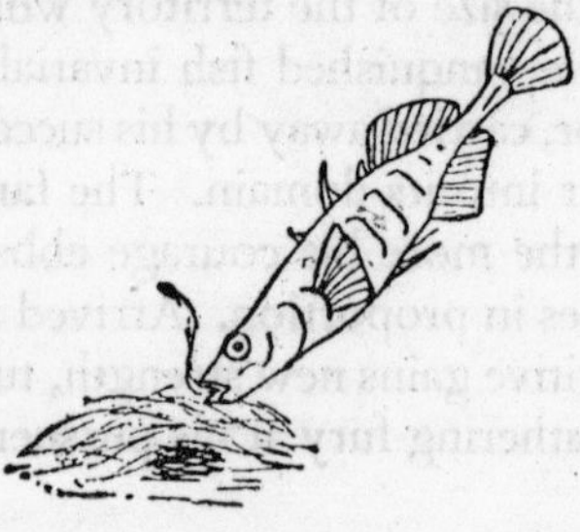

Unlike the fighting-fish, the sticklebacks do not waste time by threatening before starting to fight, but will do so after or between battles. This, in itself, implies that they never fight to a finish, although from their method of fighting, the contrary might be expected. Thrust and counter-thrust follow each other so quickly that the eye of the observer can scarcely follow them. The large ventral spine, that appears so ominous, plays in reality quite a subordinate role. In older aquarium literature it is often stated that these spines are used so effectively that one of the fighters may sink down dead, perforated by the spine of his opponent. Apparently the writers of these works have never tried to "perforate" a stickleback; for even a dead stickleback will slip from under the sharpest scalpel before one is able to penetrate its tough skin, even in places where it is not reinforced by bony armour. Place a dead stickleback on a soft surface—which certainly offers a much better resistance than water—and try to run it through with a sharp needle. You will be surprised at the force required to do so. Owing to the extreme toughness of the sticklebacks' skin, no serious wounds can be inflicted in their natural battles, which, as compared with those of the fighting-fish, are absurdly harmless. Of course, in the confined space of a small tank a stronger male stickleback may harry a weaker one to death, but rabbits and turtle-doves, in analogous conditions, will do the same thing to each other.

The stickleback and the fighting-fish are as different in love as they are in fight, yet, as parents, they have much in common. In both species it is the male and not the female that undertakes the building of the nest and the care of the young, and the future father only begins to think of love when the cradle for the expected children is ready. But here the similarities end and the differences begin. The cradle of the stickleback lies, in a manner of speaking, under the floor, that of the fighting-fish above the ceiling: that is to say, the former digs a little hollow in the bottom, and the latter builds his nest on the surface of the water; the one uses for nest construction plant strands and a special sticky kidney secretion, the other uses air and spittle. The castle-in-the-air of the fighting-fish, as also that of his nearer relations, consists of a little pile of air bubbles, stuck closely together, which protrudes somewhat over the water surface; the bubbles are coated with a tough layer of spittle and are very resistant. Already while building, the male radiates the most gorgeous colours, which gain in depth and iridescence when a female approaches. Like lightning, he shoots towards her and, glowing, halts. If the female is prepared to accept him, she demonstrates it by investing herself with a characteristic, if modest, colouring consisting of light grey vertical stripes on a brown background. With fins closely folded, she swims towards the male, who, trembling with excitement, expands all his fins to breaking point and holds himself in such a position that the dazzling brilliance of his full broadside is presented to his bride. Next moment he swims off with a sweeping, gracefully sinuous movement, in the direction of the nest. The beckoning nature of this gesture is at once apparent even when seen for the first time. The essentially ritual nature of this swimming movement is easily understood: everything that enhances its optical effect, as the sinuous movements of the body or the waving of the tail fin, is exaggerated in mimic, whereas all the means of making it mechanically effective are decreased. The movement says: "I am swimming away from you, hurry up and follow me!" At the same time, the fish swims neither fast nor far, and turns back immediately to the female, who is following but timidly and shyly in his wake.

In this way the female is enticed under the bubble nest, and now follows the wonderful love-play which resembles, in delicate grace, a minuet, but in general style the trance dance of a Balinese temple dancer. In this love-dance, by age-old law, the male must always exhibit his magnificent broadside to his partner, but the female must remain constantly at right angles to him. The male must never obtain so much as a glimpse of her flanks, otherwise he will immediately become angry and unchivalrous; for standing broadside means, in these fishes as in many others, aggressive masculinity and elicits instantaneously in every male a complete change of mood: hottest love is transformed to wildest hate. Since the male will not now leave the nest, he moves in circles round the female, and she follows his every movement by keeping her head always turned towards him; the love-dance is thus executed in a small circle, exactly under the middle of the nest. Now the colours become more glowing, more frantic the movements, ever smaller the circles, until the bodies touch. Then, suddenly, the male slings his body tightly round the female, gently turns her on her back

and, quivering, both fulfil the great act of reproduction. Ova and semen are discharged simultaneously.

The female remains for a few seconds as though benumbed, but the male has important things to attend to at once. The minute, glass-clear eggs are considerably heavier than water and sink at once to the ground. Now the posture of the bodies in spawning is such that the sinking eggs are bound to drift past the downward directed head of the male and thus catch his attention. He gently releases the female, glides downwards in pursuit of the eggs and gathers them up, one after the other, in his mouth. Turning upwards again, he blows the eggs into the nest. They now

miraculously float instead of sinking. This sudden and amazing change of density is caused by a coating of buoyant spittle in which the male has enveloped every egg while carrying it in his mouth. He has to hurry in this work, for not only would he soon be unable to find the tiny, transparent globules in the mud, but, if he should delay a second longer, the female would wake from her trance and, also swimming after the eggs, would likewise proceed to engulf them. From these actions it would appear, at first sight, that the female has the same intentions as her mate. But if we wait to see her packing the eggs in the nest, we will wait in vain, for these eggs will disappear, irrevocably swallowed. So the male knows very well why he is hurrying, and he knows, too, why he no longer allows the female near the nest when, after ten to twenty matings, all her eggs have been safely stored between the air bubbles.

The family life of the beautiful and courageous fishes of the cichlid group is much more highly developed than that of the fighting-fish. Here both male and female care for their young, which follow their parents as chickens the hen. For the first time in the ascending ranks of the scale of living creatures, we see in these cichlids a type of behaviour which human beings consider highly moral: male and female remain in close connubial partnership even after reproduction is completed. And not only do they remain so as long as the care of the brood necessitates it, but—and this is what counts—still longer. It is usually described as "marriage" when both partners together fend for the brood, though, for this purpose, no really personal ties need exist between male and female; but in cichlids they do exist.

In order to ascertain objectively whether an animal recognizes its mate personally, the latter must be substituted, in experiment, by another of the same sex and in exactly the same phase of the reproductive cycle. If, for instance, in a pair of birds just beginning to nest, we replace the female by one that is already in the psycho-physiological phase of feeding its young, its instinctive behaviour will naturally fail to harmonize with that of the male. If the male then reacts inimically it is impossible to say whether he really notices that the substituted female is not his wife or whether it merely annoys him that she behaves "wrongly". I was greatly interested to find out how cichlids—the only fish that live in a "life-time marriage"—behave in this respect. The first thing necessary for the elucidation of this question was the possession of two pairs in exactly the same stages of their reproductive cycles. I was lucky enough, in the year 1941, to have two pairs of the magnificent South American cichlid *Herichthys cyanoguttatus*, which fulfilled this condition. The Latin name, which, translated into English, means "Blue-spotted Hero-Fish", is apposite: on a velvet-black background, deep turquoise-blue iridescent spots form an intricate mosaic, and a breeding-pair of these fishes displays, even to the largest adversary, a heroism which justifies the second part of their name. When I first got them, my five young fishes of this species were neither blue-spotted nor heroic. After some weeks of concentrated feeding in a large sunny aquarium they grew and flourished and one day one of the two biggest fishes showed his nuptial colours. He took possession of the left-hand lower front cover of the container, hollowed out a deep nesting cavity, and began to prepare a large smooth stone for spawning by carefully freeing it from algæ and other deposits. The other four fish stood huddled in an anxious group in the right upper rear corner. But by the next morning one of them, a smaller one, had also put on its gala dress; the velvet-black breast, devoid of blue spots, proved it to be a female. The male proceeded straightway to fetch his lady home, by a ceremony very similar to that described in the fighting-fish.

The pair now stood over the nesting-place and defended its area valorously. This was no laughing matter for the three remaining fish, who were allowed no rest, being chased to and fro all the time, and it says much for the name-giving heroism of the species that after some days the second largest male plucked up enough courage to make conquest of the opposite corner. The two males now sat facing each other like two hostile knights in their castles. The border lay nearer the castle of the second one, a fact which will be appreciated after what I have said about territorial fighting: the fighting potential of the single male was smaller than the combined forces of the pair, and his territory was correspondingly smaller. The solitary male, which we will simply call number two, sallied forth again and again from his castle with the intention of abducting his neighbour's wife. His attempts, however, were fruitless and brought him nothing but discomfiture. Every time he tried to pay court to her by displaying his magnificent broadside, she repaid his efforts by a ramming thrust in his unprotected flank. This situation remained unaltered for several days; then a second female donned her bridal dress and a happy end seemed imminent. But nothing of the kind occurred. On the contrary, the newly matured female paid as little attention to male number two as he to her, and each ignored the other completely. Female number two tried again and again to approach male number one. Every time he swam towards his home, she followed, in the attitude of a female being led to the nest. She "considered" herself as being enticed nestwards whenever male number one, after a sally, swam back in that direction. His wife seemed to grasp the situation thoroughly, judging by the ferocity with which she attacked the intruder every time she approached; in this her husband only mildly participated. Male and female number two just did not exist for each other and each of them had eyes only for the opposite sex in the happily married pair which showed so little interest in them.

This situation would have lasted long if I had not intervened and put the number twos in another, identical aquarium. Separated from the objects of their unrequited love, the two quickly found solace in each other and became a pair.

After a few days the two pairs spawned in the same hour. Now I had exactly what I had wanted; namely, two cichlid pairs of the same species in the identical phase of reproduction. As the breeding of these fish, at that time rare, meant much to me, I waited with my experiment till the young of both couples were big enough to exist independently even in the event of a complete marital rupture of their parents.

Then I exchanged the females. The result was ambiguous and gave no definite answer to the question as to whether the fish knows his own mate personally. My interpretation of what now followed will be considered by many as daring, and it certainly needs further experimental corroboration.

Male number two accepted female number one as soon as she was placed with him. But it did not appear to me as though he was unaware of the difference, indeed his movements at the "changing-of-the-guard" ceremony, and whenever he met his wife, seemed to have increased in fire and vigour. The female immediately acquiesced in the ceremonies of the male and adapted herself without demur to her role. This, however, did not mean much, because in this phase the female is only occupied with the young and has little interest in the male.

The proceedings in the other aquarium, in which I had introduced female number two to male number one and his offspring, took an entirely different turn. Here, too, the female was only interested in the children, swam immediately to the shoal, and, herself upset by the change, began anxiously to gather the young ones about her. This is just what female number one had done in the other aquarium. But the contrast lay in the behaviour of the males; while male two had received the substitution of the new female with friendly glowing ceremonies, male one remained suspiciously guarding his flock, refused to let the female relieve him of his charge and in the next moment attacked her with a furious ram-thrust. At once, some silvery scales danced like sunbeams to the bottom of the tank and I had to interfere with alacrity in order to rescue the female, who otherwise would certainly have been gored to death.

What had happened? The fish which had received the "prettier" female, the one to whom he had previously paid court, was quite content with the exchange, but the other, who had been landed with the formerly rejected female in place of his wife, was, not unjustifiably, furious and now attacked her much more relentlessly than he had done at first, in the presence of his wife. I am convinced that male number two, who had received an improvement on his wife, noticed the difference too.

Almost more interesting and, for the observer, more fascinating than the sexual behaviour of these fishes is their method of caring for their brood. Anyone who has watched their behaviour, as they fan a continuous stream of fresh water towards their eggs or small babies lying in the nest, or as, with military exactitude, they relieve each other of duty, or as later, when the brood has learned to swim, they lead them carefully through the water, will never forget these scenes. The prettiest sight of all is when the children which can already swim are put to bed in the evening. For every evening, until they reach the age of several weeks, the young are brought, as dusk falls, back to the nesting hollow where they spent their earliest childhood. The mother stands above the nest and gathers the young about her. This she does by certain signal movements of her fins.

These details of behaviour are particularly clearly developed in the gorgeous jewel fish (*Hemichromis bimaculatus*), one of the most beautiful of all cichlids. I think Rupert Brooke must have been thinking of this species when he wrote the lines:

> Red darkness of the heart of roses,
> Blue brilliant from dead starless skies,
> And gold that lies behind the eyes,
>
> Lustreless purple, hooded green,
> The myriad hues that lie between
> Darkness and darkness!

The iridescent, brilliant blue spots in the red darkness of the dorsal fin play a special role when the female jewel fish is putting her babies to bed. She jerks her fin rapidly up and down, making the jewels flash like a heliograph. At this, the young congregate under the mother and obediently descend into the nesting-hole. The father, in the meantime, searches the whole tank for stragglers. He does not coax them along, but simply inhales them into his roomy mouth, swims to the nest, and blows them into the hollow. The baby sinks at once heavily to the bottom and remains lying there. By an ingenious arrangement of reflexes, the swim-bladders of young "sleeping" cichlids contract so strongly that the tiny fishes become much heavier than water and remain, like little stones, lying in the hollow, just as they did in their earliest childhood before their swim-bladder was filled with gas. The same reaction of "becoming heavy" is also elicited when a parent fish takes a young one in its mouth. Without this reflex mechanism it would be impossible for the father, when he gathers up his children in the evening, to keep them together.

I once saw a jewel fish, during such an evening transport of strayed children, perform a deed which absolutely astonished me. I came, late one evening, into the laboratory. It was already dusk and I wished hurriedly to feed a few fishes which had not received anything to eat that day; amongst them was a pair of jewel fishes who were tending their young. As I approached the container, I saw that most of the young were already in the nesting hollow, over which the mother was hovering. She refused to come for the food when I threw pieces of earthworm into the tank. The father, however, who, in great excitement, was dashing backwards and forwards searching for truants, allowed himself to be diverted from his duty by a nice hind-end of earthworm (for some unknown reason this end is preferred by all worm-eaters to the front one). He swam up and seized the worm, but, owing to its size, was unable to swallow it. As he was in the act of chewing this mouthful, he saw a baby fish swimming by itself across the tank; he started as though stung, raced after the baby and took it into his already filled mouth. It was a thrilling moment. The fish had in its mouth two different things, of which one must go into the stomach and the other into the nest. What would he do? I must confess that, at that moment, I would not have given twopence for the life of that tiny jewel fish. But wonderful what really happened! The fish stood stock still with full cheeks, but did not chew. If ever I have seen a fish think, it was in that moment! What a truly remarkable thing that a fish can find itself in a genuine conflicting situation and, in this case, behave exactly as a human being would; that is to say, it stops, blocked in all directions, and can go neither forward nor backward. For many seconds the father jewel fish stood riveted, and one could almost see how his feelings were working. Then he solved the conflict in a way for which one was bound to feel admiration: he spat out the whole contents of his mouth: the worm fell to the bottom, and the little jewel fish, becoming heavy in the way described above, did the same. Then the father turned resolutely to the worm and ate it up, without haste but all the time with one eye on the child which "obediently" lay on the bottom beneath him. When he had finished, he inhaled the baby and carried it home to its mother.

Some students, who had witnessed the whole scene, started as one man to applaud.

5. LAUGHING AT ANIMALS

IT is seldom that I laugh at an animal, and when I do, I usually find out afterwards that it is at myself, at the human being whom the animal has portrayed in a more or less pitiless caricature, that I have laughed. We stand before the monkey house and laugh, but we do not laugh at the sight of a caterpillar or a snail, and when the courtship antics of a lusty greylag gander seem so incredibly funny, it is only because our human youth behaves in a very similar fashion.

The initiated observer seldom laughs at the bizarre in animals. It often annoys me when visitors at a zoo or aquarium laugh at an animal that, in the course of its evolutionary adaptation, has developed a body-form which now deviates from the usual. The public is then deriding things which, to me, are holy: the riddles of the Genesis, the Creation and the Creator. The grotesque forms of a chameleon, a puffer or an anteater awake in me feelings of awed wonder, but not of amusement.

Of course I have laughed at unexpected drollness, although such amusement is in itself not less stupid than that of the public that annoys me. When the queer, land-climbing fish Periophthalmus was first sent to me and I saw how one of these creatures leaped, not out of the water-basin, but on to its edge and, raising its head with its pug-like face towards me, sat there perched, staring at me with its goggling, piercing eyes, then I laughed heartily. Can you imagine what it is like when a fish, a real and unmistakable vertebrate fish, first of all sits on a perch, like a canary, then turns its head towards you like a higher terrestrial animal, like anything but a fish, and then, to crown all, fixes you with a binocular stare? This same stare gives the owl its characteristic and proverbially wise expression, because, even in a bird, the two-eyed gaze is unexpected. But here, too, the humour lies more in the caricature of the human than in the actual drollness of the animal.

In the study of the behaviour of the higher animals, very funny situations are apt to arise, but it is inevitably the observer, and not the animal, that plays the comical part.

The comparative ethologist's method in dealing with the most intelligent birds and mammals often necessitates a complete neglect of the dignity usually to be expected in a scientist. Indeed, the uninitiated, watching the student of behaviour in operation, often cannot be blamed for thinking that there is madness in his method. It is only my reputation for harmlessness, shared with the other village idiot, which has saved me from the mental home. But in defence of the villagers of Altenberg I must recount a few little stories.

I was experimenting at one time with young mallards to

find out why artificially incubated and freshly hatched ducklings of this species, in contrast to similarly treated greylag goslings, are unapproachable and shy. Greylag goslings unquestioningly accept the first living being whom they meet as their mother, and run confidently after him. Mallards, on the contrary, always refused to do this. If I took from the incubator freshly hatched mallards, they invariably ran away from me and pressed themselves in the nearest dark corner. Why? I remembered that I had once let a muscovy duck hatch a clutch of mallard eggs and that the tiny mallards had also failed to accept this foster-mother. As soon as they were dry they had simply run away from her, and I had trouble enough to catch these crying, erring children. On the other hand, I once let a fat white farmyard duck hatch out mallards, and the little wild things ran just as happily after her as if she had been their real mother. The secret must have lain in her call-note, for in external appearance the domestic duck was quite as different from a mallard as was the muscovy; but what she had in common with the mallard (which, of course, is the wild progenitor of our farmyard duck) were her vocal expressions. Though, in the process of domestication, the duck has altered considerably in colour-pattern and body-form, its voice has remained practically the same. The inference was clear: I must quack like a mother mallard in order to make the little ducks run after me. No sooner said than done. When, one Whit-Sunday, a brood of pure-bred young mallards was due to hatch, I put the eggs in the incubator, took the babies, as soon as they were dry, under my personal care, and quacked for them the mother's call-note in my best Mallardese. For hours on end I kept it up, for half the day. The quacking was successful. The little ducks lifted their gaze confidently towards me, obviously had no fear of me this time, and as, still quacking, I drew slowly away from them, they also set themselves obediently in motion and scuttled after me in a tightly huddled group, just as ducklings follow their mother. My theory was indisputably proved. The freshly hatched ducklings have an inborn reaction to the call-note, but not to the optical picture of the mother. Anything that emits the right quack note will be considered as mother, whether it is a fat white Pekin duck or a still fatter man. However, the substituted object must not exceed a certain height. At the beginning of these experiments, I had sat myself down in the grass amongst the ducklings and, in order to make them follow me, had dragged myself, sitting, away from them. As soon, however, as I stood up and tried, in a standing posture, to lead them on, they gave up, peered searchingly on all sides, but not upwards towards me, and it was not long before they began that penetrating piping of abandoned ducklings that we are accustomed simply to call "crying". They were unable to adapt themselves to the fact that their foster-mother had become so tall. So I was forced to move along, squatting low, if I wished them to follow me. This was not very comfortable; still less comfortable was the fact that the mallard mother quacks unintermittently. If I ceased for even the space of half a minute from my melodious "Quahg, gegegegeg, Quahg, gegegegeg", the necks of the ducklings became longer and longer, corresponding exactly to "long faces" in human children—and did I then not immediately recommence quacking, the shrill weeping began anew. As soon as I was silent, they seemed to think that I had died, or perhaps that I loved them no more: cause enough for crying! The ducklings, in contrast to the greylag goslings, were most demanding and tiring charges, for imagine a two-hour walk with such children, all the time squatting low and quacking without interruption! In the interests of science I submitted myself literally for hours on end to this ordeal. So it came about, on a certain Whit-Sunday, that, in company with my ducklings, I was wandering about, squatting and quacking, in a May-green meadow at the upper part of our garden. I was congratulating myself on the obedience and exactitude with which my ducklings came waddling after me, when I suddenly looked up and saw the garden fence framed by a row of dead-white faces: a group

of tourists was standing at the fence and staring horrified in my direction. Forgivable! For all they could see was a big man with a beard dragging himself, crouching, round the meadow, in figures of eight, glancing constantly over his shoulder and quacking—but the ducklings, the all-revealing and all-explaining ducklings, were hidden in the tall spring grass from the view of the astonished crowd.

As I shall tell in a later chapter, jackdaws long remember someone who has laid hands on them and thereby elicited a "rattling" reaction. Therein lay a considerable impediment to the ringing of the young jackdaws reared in my colony. When I took them out of the nest to mark them with aluminium rings, I could not help the older jackdaws seeing me and at once raising their voices to a wild rattling concert. How was I to stop the birds developing a permanent shyness for me as a result of the ringing procedure, a state of affairs which would have been immeasurably detrimental to my work? The solution was obvious: disguise. But what? Again quite easy. It lay ready to hand in a box in the loft and was very well suited for my purpose, although, normally, it was only brought out every sixth of December to celebrate the old Austrian festival of St Nicholas and the Devil. It was a gorgeous, black, furry devil's costume with a mask covering the whole head, complete with horns and tongue, and a long devil's tail which stuck well out from the body. I wonder what you would think if, on a beautiful June day, you suddenly heard from the gabled roof of a high house, a wild rattling noise and, looking up, you saw Satan himself, equipped with horns, tail and claws, his tongue hanging out with the heat, climbing from chimney to chimney, surrounded by a swarm of black birds making ear-splitting rattling cries. I think this whole alarming impression disguised the fact that the devil was fixing, by means of a forceps, aluminium rings to the legs of young jackdaws, and then replacing the birds carefully in their nests. When I had finished the ringing, I saw for the first time that a large crowd of people had collected in the village street, and were

looking up with expressions just as aghast as those of the tourists at the garden fence. As I would have defeated my own object by now disclosing my identity, I just gave a friendly wag of my devil's tail and disappeared through the trap-door of the loft.

The third time that I was in danger of being delivered up to the psychiatric clinic was the fault of my big yellow-crested cockatoo Koka. I had bought this beautiful and very tame bird shortly before Easter, for a considerable sum of money. It was many weeks before the poor fellow had overcome the mental disturbances caused by his long imprisonment. At first he could not realize that he was no longer fettered and could now move about freely. It was a pitiable sight to see this proud creature sitting on a branch, ever and anon preparing himself for flight, but not daring to take off, because he could not believe that he was no longer on the chain. When at last he had overcome this inward resistance he became a lively and exuberant being and developed a strong attachment for my person. As soon as he was let out of the room in which we still shut him up at night-time, he flew straight off to find me, displaying thereby an astonishing intelligence. In quite a short time he realized where I was probably to be found. At first he flew to my bedroom window, and, if I was not there, down to the duck-pond; in short, he visited all the sites of my morning inspection at the various animal pens in our research station. This determined quest was not without danger to the cockatoo because, if he failed to find me, he extended his search farther and farther, and had several times lost his way on such occasions. Accordingly, my fellow-workers had strict instructions not to let the bird out during my absence.

One Saturday in June, I got off the train from Vienna at Altenberg station, in the midst of a gathering of bathers, such as often flock to our village at fine week-ends. I had gone only a few steps along the street and the crowd had not yet dispersed when, high above me in the air, I saw a bird whose species I could not at first determine. It flew with slow, measured wing-beats, varied at set intervals by longer periods of gliding. It seemed too heavy to be a buzzard; for a stork, it was not big enough and, even at that height, neck and feet should have been visible. Then the bird gave a sudden swerve so that the setting sun shone for a second full on the underside of the great wings, which lit up like stars in the blue of the skies. The bird was white. By Heaven, it was my cockatoo! The steady movements of his wings clearly indicated that he was setting out on a long-distance flight. What should I do? Should I call to the bird? Well, have you ever heard the flight-call of the greater yellow-crested cockatoo? No? But you have probably heard pig-killing after the old method. Imagine pig-squealing at its most voluminous taken up by a microphone and magnified many times by a good loudspeaker. A man can imitate it quite successfully, though somewhat feebly, by bellowing at the top of his voice "O-ah". I had already proved that the cockatoo understood this imitation and promptly "came to heel". But would it work at such a height? A bird always has great difficulty in making the decision to fly downwards at a steep angle. To yell, or not to yell, that was the question. If I yelled and the bird came down, all would be well, but what if it sailed calmly on through the clouds? How would I then explain my song to the crowd of people? Finally, I did yell. The people around me stood still, rooted to the spot. The bird hesitated for a moment on outstretched wings, then, folding them, it descended in one dive and landed upon my outstretched arm. Once again I was master of the situation.

On another occasion, the frolics of this bird gave me quite a serious fright. My father, by that time an old man, used to take his siesta at the foot of a terrace on the south-west side of our house. For medical reasons, I was never quite happy to think of him exposed to the glaring midday sun, but he would let nobody break him of his old habit. One day, at his siesta time, I heard him, from his accustomed place, swearing like a trooper, and as I raced round the corner of the house, I saw the old gentleman swaying up the drive in a cramped position, bending forwards, his arms tightly folded about his waist. "In heaven's name, are you ill?" "No," came the embittered response, "I am not ill, but that confounded creature has bitten all the buttons off my trousers while I was fast asleep!" And that is what had happened. Eye-witnesses at the scene of the crime discovered, laid out in buttons, the whole outline of the old professor: here the arms, there the waistcoat, and here, unmistakably, the buttons off his trousers.

One of the nicest cockatoo-tricks was one which, in fanciful inventiveness, equalled the experiments of monkey or human children. It arose from the ardent love of the bird for my mother, who, so long as she stayed in the garden in summer-time, knitted without stopping. The cockatoo seemed to understand exactly how the soft skeins worked and what the wool was for. He always seized the free end of the wool with his beak and then flew lustily into the air, unravelling the ball behind him. Like a paper kite with a long tail, he climbed high and then flew in regular circles round the great lime-tree which stood in front of our house. Once, when nobody was there to stop him, he encircled the tree, right up to its summit, with brightly coloured woollen strands which it was impossible to disentangle from the wide-spreading foliage. Our visitors used to stand in mute astonishment before this tree, and were unable to understand how and why it had been thus decorated.

The cockatoo paid court to my mother in a very charming way, dancing round her in the most grotesque fashion, folding and unfolding his beautiful crest and following her wherever she went. If she were not there, he sought her just as assiduously as he had been used, in his early days, to search for me. Now my mother had no less than four sisters. One day these aunts, in company with some equally aged ladies of their acquaintance, were partaking of tea in the veranda of our house. They sat at a huge round table, a plate of luscious home-grown strawberries in front of each, and in the middle of the table a large, very shallow bowl of finest icing sugar. The cockatoo, who was flying accidentally or wittingly past, espied, from without, my mother who was presiding at this festive board. The next moment, with a perilous dive, he steered himself through the doorway, which, though wide, was nevertheless narrower than the span of his wings. He intended to land before my mother on the table where he was accustomed to sit and keep her company while she knitted; but this time he found the runway encumbered with numerous obstacles to flying technique and, into the bargain, he was in the midst of unknown faces. He considered the situation, pulled himself up abruptly in mid-air, hovering over the table like a helicopter, then, turning on his own axis, he opened the throttle again and the next second had disappeared. So also had the icing sugar from the shallow bowl, out of which the propeller wind had wafted every grain. And around the table sat seven powdered ladies, seven rococo ladies whose faces, like lepers', were white as snow and who held their eyes tight shut. Beautiful!

6. PITYING ANIMALS

Pity—best taught by fellowship of woe.

COLERIDGE

IF one listens to the remarks of the visitors in any of the larger zoos one will frequently find that people are in the habit of wasting sentimental pity on animals that are absolutely contented with their lot, while the genuine suffering which is to be found in every zoological gardens may pass unnoticed. People are specially apt to pity those animals which, owing to their particular emotional associations, play a prominent role in literature, like the nightingale, the lion or the eagle.

How much the real character of the singing nightingale is generally misunderstood is shown by the fact that in literature the bird is always represented as a female; in the German language the very word "nightingale" is of female gender. It is, of course, the male that sings, and the meaning of the song is a defensive threat to all other males that might invade the songster's territory, as well as an invitation to any passing female to join him. To anybody really familiar with birds, the masculinity of the singing nightingale is so blatantly apparent that to attribute loud song to a female bird is as comically incongruous as it would be to the student of literature had Tennyson invested Guinevere with a beard. For this reason I was never able to appreciate Oscar Wilde's beautiful fairy-tale of the nightingale who made the red rose "of music by moonlight" and "stained it with her own heart's blood", and I must confess that I was heartily relieved when at last, with the thorn in her heart, the virago ceased her lusty singing.

Later on I shall deal with the supposed suffering of caged birds. Of course, the singly kept male nightingale may suffer some sort of disappointment when despite his prolonged singing no female puts in an appearance, but, owing to the excess of males, this is also liable to happen in nature.

The lion is another animal very often misrepresented in literature, both as to habitat and to character. The English call him King of the Jungle—thus relegating him to much too wet a locality, while the Germans, with customary thoroughness, go to the other climatic extreme and deposit him in the desert, calling him "Wüstenkönig" (Desert King). In reality, he prefers the happy medium and lives in steppes or savannahs. His majesty of bearing, to which he owes the first part of his title, is due to the simple fact that, being a hunter of large animals of the open plains, he habitually surveys the far distance and disregards everything moving in the foreground.

The lion suffers less under close confinement than most other carnivores of equal mental development, for the simple reason that he has a lesser urge for movement. To put it crudely, the lion is about the laziest of the predatory beasts: he is indeed quite enviably indolent. Under natural conditions he covers enormous distances, but obviously only under pressure of hunger and not from any inward drive. Therefore, it is seldom that a lion in captivity is seen pacing restlessly to and fro in his cage, as wolves and foxes will do for hours at a stretch. If the pent-up drive for locomotion urges him, for once, to walk up and down the length of his cage, his movements bear the character of a leisurely after-dinner stroll and have nothing of the frantic haste with which captive canine carnivores discharge their frustrated urge to cover large distances. In the Berlin Zoo, a huge paddock with desert sand and yellow rocky crags was made for the lions, but this expensive construction proved very nearly useless; a gigantic model with stuffed beasts might have served the same purpose, so lazily did the lions lie about in their romantic surroundings.

And now for the eagle! I hate to shatter the fabulous illusions about this glorious bird, but I must adhere to the truth: all true birds of prey are, compared with passerines or parrots, extremely stupid creatures. This applies particularly to the golden eagle, "the eagle" of our mountains and our poets, which is one of the most stupid among them, much more so indeed than any barnyard fowl. This, of course, does not preclude this proud bird from being beautiful and impressive and embodying the very essence of wild life; but here we have to deal with the mental qualities of the creature, its supposed love of freedom and its imaginary suffering in captivity.

I still remember what disappointment was caused me by my first and only eagle, an imperial eagle which I bought, out of pity, from a wandering menagerie. She was a beautiful female bird, nearly matured in colour, a sign that she boasted several years. She was completely tame and greeted her keeper, and later myself, with a curious gesture of affection in which she turned her head upside down, so that the fearful curve of her beak pointed perpendicularly upwards. At the same time, the creature spoke in tones so quiet and docile as to be worthy of a turtle-dove; moreover, compared with the latter, she was a veritable lamb (see

Chapter 12). I originally bought her because I intended to train her for hawking, as many Asiatic peoples are known to do with eagles. I did not flatter myself that I should acquire any particular success in this noble sport, but I hoped, if only by using a domestic rabbit as bait, to make observations on the hunting behaviour of one of these large birds of prey. This plan failed because my eagle, even when she was hungry, refused to harm a hair of the rabbit's body.

She also showed little inclination for flight, although she was healthy and strong and possessed excellent wing feathers. Ravens, cockatoos and buzzards fly for the pleasure of it, and playfully enjoy the fullness of their ability. But not so this eagle. She only flew at all when there were favourable up-currents over our garden which enabled her to soar without expending much muscle power, and even then she never circled really high. When she wished to come down, she invariably failed to find her way home. She circled around without any sense of direction and landed at last somewhere in the neighbourhood. There she sat unhappy and benighted and waited till I fetched her. Perhaps she might have come home alone, but she was so conspicuous that somebody always telephoned from somewhere that the eagle was sitting on this or that roof where a crowd of small boys was pelting her with stones. Then I had to go there on foot, because the silly creature was desperately afraid of a bicycle. Again and again have I plodded in this manner wearily home, the eagle on my arm. In the end, because I did not wish to keep her permanently chained up, I gave her to the Schönbrunn Zoo.

The large aviaries that are to be found to-day in every good zoo allow adequately for the flying inclinations of an eagle, and could one interrogate one of these birds I think it would submit the following information on the subject of its wishes or complaints: "We suffer here mostly from overpopulation of our enclosure. As often as I or my wife add a twig to our half-finished nest, one of these vile white-headed vultures comes and carries it away. The company of the American bald-headed eagles also gets on my nerves; they are stronger than we are and horribly overbearing. And still worse is the Andean condor, that disagreeable fellow. The food is quite good, but we get a bit too much horseflesh; I should far prefer smaller animals, for example rabbits, including their fur and bones." The eagle would not speak of a longing for golden liberty.

Now which are the animals really to be pitied in captivity? I have already given a partial answer to this question: In the first place, those clever and highly developed beings whose lively mentality and urge for activity can find no outlet behind the bars of the cage. Furthermore, all those animals which are ruled by strong drives that cannot be satisfied in captivity. This is most conspicuous, even for the uninitiated, in the case of animals which, when living in a free state, are accustomed to roam about widely and therefore have a correspondingly strong drive for locomotion. Owing to this frustrated desire, foxes and wolves housed, in many old-fashioned zoos, in cages which are far too small, are among the most pitiable of all caged animals.

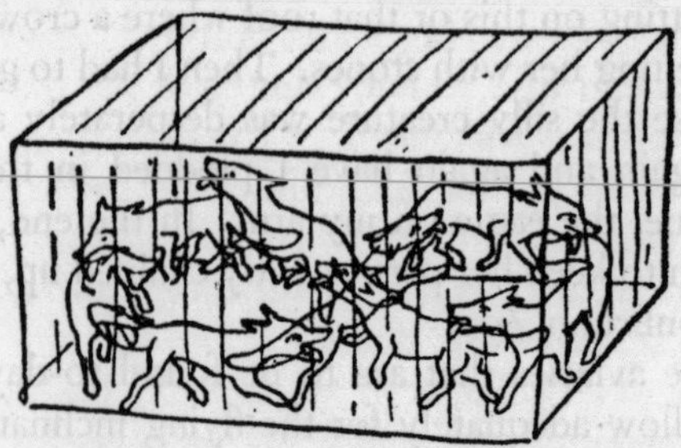

Another piteous scene, seldom noticed by ordinary zoo visitors, is enacted by some species of swans at migration time. These creatures, like most other water-fowl, are in zoos generally rendered incapable of flight by the operation of "pinioning", that is the amputation of the wing bone at the metacarpal joint. The birds never really grasp that they can fly no more and they try again and again. I do not like pinioned water-birds; the missing tip of one wing and the still sadder picture that the bird makes when it spreads its wings spoil most of my pleasure in it, even if it belongs to one of those species which do not suffer mentally by the mutilation.

Though pinioned swans generally seem happy and signify their contentment, under proper care, by hatching and rearing their young without any trouble, at migration time things become different: the swans repeatedly swim to the lee side of the pond, in order to have the whole extent of its surface at their disposal when trying to take off against

the wind. All the while, their sonorous flying calls can be heard as they try to rise, and again and again the grand preparations end in a pathetic flutter of the one and a half wings; a truly sorry picture!

But of all animals that suffer under the inefficient methods of many zoological gardens, by far the most unfortunate are those mentally alert creatures of whom we have spoken above. These, however, rarely awaken the pity of the zoo visitor, least of all when such an originally highly intelligent animal has deteriorated, under the influence of close confinement, into a crazy idiot, a very caricature of its former self. I have never heard an exclamation of sympathy from the onlookers in the parrot house. Sentimental old ladies, the fanatical sponsors of the societies for prevention of cruelty to animals, have no compunction in keeping a grey parrot or cockatoo in a relatively small cage or even chained to a perch. Now these larger species of the parrot tribe are not only clever but mentally and bodily uncommonly vivacious; and, together with the large corvines, they are probably the only birds which can suffer from that state of mind, common to human prisoners, called boredom. But nobody pities these pathetic creatures in their bell-shaped cages of martyrdom. Uncomprehendingly, the fond owner imagines that the bird is bowing when it constantly repeats the bobbing head movements which, in reality, are the stereotyped remnants of its desperate attempts to escape from its cage. Free such an unhappy prisoner, and it will take weeks, even months, before it really dares to fly.

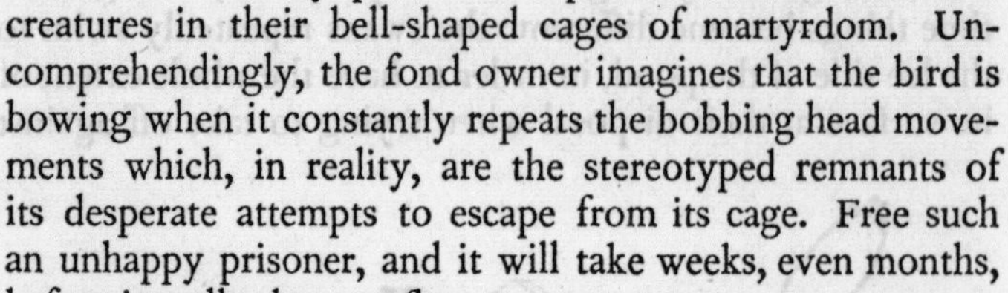

More wretched still in confinement are monkeys, above all the anthropoid apes. They are the only captive animals which can derive serious bodily harm from their mental suffering. Anthropoid apes can become literally bored to death, particularly when they are kept alone in too small cages. For this and no other reason, it is easily explained why monkey babies thrive admirably in private ownership where they "live as the family", but immediately begin to pine when, having become too large and dangerous, they are transferred to the cages of the nearest zoo. My capuchin monkey Gloria was overtaken by this fate. It is no exaggeration when I say that real success in the keeping of anthropoid apes was only achieved when it was realized how to prevent the mental

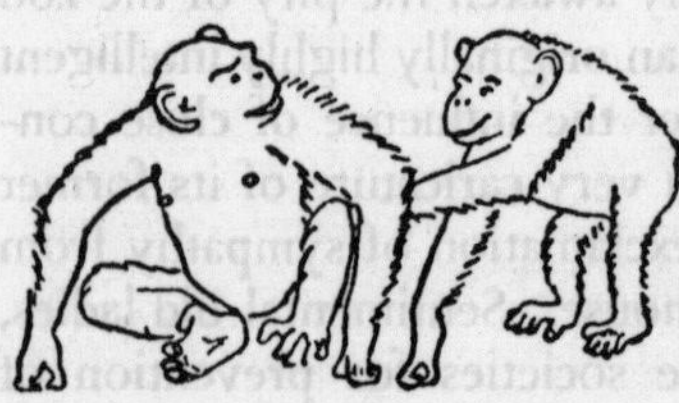

sufferings caused by confinement. I have beside me the wonderful chimpanzee book by Robert Yerkes, one of the best authorities on this kind of ape; from this work it may be concluded that mental hygiene plays just as important a role as physical, in the maintenance of health of these most human of all animals. On the other hand, to keep anthropoid apes in solitary confinement and in such small cages as are still to be found in many zoos, is an act of cruelty which should be punishable by law.

In his big anthropoid ape station in Orange Park, Florida, Yerkes has kept, for many years, a chimpanzee colony, which has multiplied freely and in which the apes live as happily as do my lesser whitethroats in their aviary, and much more happily than you or I.

7. BUYING ANIMALS[1]

Brothers and Sisters, I bid you beware
Of giving your heart to a dog to tear.

RUDYARD KIPLING

FEW people know which animals are suitable and responsive objects of care. Again and again nature-loving people attempt to keep house pets, and again and again the attempt fails because of inadequate technique and wrong selection of the animal. Moreover, most of our animal dealers are incapable of assessing a customer and of advising him in his choice.

The beginner must first make up his mind what he really expects from his charge. The wish to keep an animal usually arises from a general longing for a bond with nature. Every animal is a piece of nature, but not every one is a suitable representative of nature to live in your house. The animals which you should not acquire can be divided into two

[1] *Publishers' note.* English law states that many of the birds mentioned in this chapter (including bullfinch, starling, siskin, goldfinch, hawfinch, chaffinch, robin, blackbird, thrush, nightingale, bearded tit, little owl, quail, sand martin and dabchick) may not be bought or sold in Great Britain, though most (except the nightingale and goldfinch) may legally be captured during the open season. However, in view of the general interest of this chapter, references to all these birds have been left in.

groups: those that cannot live with you and those with which you cannot live. To the first group belong those sensitive creatures which are hard to maintain in a state of health, and the second includes most of those animals of which I have already spoken in the chapter "Animals as a Nuisance". A considerable part of those which we can buy in a pet shop belong to one or other of the two groups. And of the rest, which are neither too delicate, nor provocative of too much annoyance to the owner, the greater part is so uninteresting that the cost of buying and the trouble of upkeep are hardly worth while. In particular, the usual house and nursery pets, such as goldfish, tortoises, canaries, guinea-pigs, caged parrots, Angora cats, lap dogs and others, are dull animals and can offer very few of those things which I am trying to impress upon the reader. Let us exclude all these from our consideration and concentrate on really interesting pets. Our choice will now depend on several other factors: How good or bad are our nerves in respect of noise? For how long are we at home every day, and at what times? Do we want simply to bring into our home a little piece of nature which charmingly warns and reminds us that the world does not consist only of asphalt, concrete and gas piping? Do we wish to fill a few square inches of our field of vision with something not made by the hand of man? Or do we want an animal for a companion?

If your eye is merely longing for a patch of natural, growing verdure and for the beauty of living things, then get an aquarium. Should you wish pleasantly to enliven your flat, then choose a pair of small birds: you have no idea how much homeliness a big cage with a happily married bullfinch pair spreads around it. The quiet, husky yet sweet song of the male bullfinch is wonderfully soothing; his dignified, measured and even polite courtship and his gentlemanly consideration for his little wife are amongst the prettiest things a bird-cage has to offer. With the welfare of these birds, you are only occupied for a few minutes daily. Their bird-seed costs but a few pence and the bit of greenstuff, which is the only variation in their menu, is easily obtained.

If, however, you want a personal contact, if you are a lonely person and want, like Byron, "to know there is an eye will mark your coming and look brighter when you come", then choose a dog. Do not think it is cruel to keep a dog in a town flat. His happiness depends largely upon how much time you can spend with him and upon how often he may accompany you on an errand. He does not mind waiting for hours at your study door if he is finally rewarded by a ten minutes' walk at your side. Personal friendship means everything to a dog; but remember, it entails no small responsibility, for a dog is not a servant to whom you can easily give notice. And remember, too, if you are an over-sensitive person, that the life of your friend is much shorter than your own and a sad parting, after ten or fifteen years, is inevitable.

If you are worried by such considerations, you can find many other creatures of lower mental development which are less "expensive" from an emotional point of view, and yet are "something to love": for instance, that most easily kept of our indigenous birds, the starling. An extraordinarily understanding friend used to describe him as "the poor man's dog". That is entirely appropriate. He has a point of character in common with the dog, namely, that he cannot be bought "ready-made". It is seldom that a dog, bought as an adult, becomes really your dog, just as seldom as your child is really your child if you, a rich man or woman, leave its upbringing to a nurse, governess or house-tutor. It is the intimate personal contact that counts. So you must feed and clean your nestling yourself, if you want a really affectionate bird of this species. The necessary trouble only lasts a short time. A young starling needs for its development, from its hatching till it is independent, only about twenty-four days. If you take it at the age of about two weeks from the nest, it is early enough and the whole rearing process takes a bare fortnight. It is not too troublesome and demands no more than that you should, with the aid of a forceps, cram food, five or six times daily, into the greedily gaping yellow throat of the nestling, and, with the same instrument, remove the droppings from the other end. These are neatly encapsulated by a thick skin which prevents them from smearing. In this way, the artificial nest always remains clean and no new "nappies" are required. You make the nest of hay and accommodate it in a little box, half shut and turned on its side so that the only opening is an aperture at the front through which you may introduce your hand; this resembles most closely a natural nesting cavity. In such a cradle the young starling always deposits its excreta towards the light so that dirt never falls into the nest, even if you are not there to remove it. Failing more natural food, raw meat or heart, bread soaked in milk, and a little chopped egg will suffice as nourishment; the addition of a little earth has a good effect. If they are obtainable, earthworms or fresh ant's eggs are a better food, being more natural. The starling requires this costly nourishment only in its infancy; as soon as it can eat by itself, it may be fed on almost any household scraps. As a staple diet for mature starlings, slightly damped wheat bran, with some crushed hemp- or poppy-seeds, is much to be recommended, since, with this type of food, the droppings are dry and almost odourless. A layer of peat moss in the drawer of the cage obviates any bird smell, even in the smallest room.

Should a starling seem too large and demanding of too much space, let me recommend you a siskin. This small bird is content with a very modest cage, requires no specially prepared food and will yet satisfy your craving for companionship. Of all the small birds I know, it is the only one which, even when captured in maturity, becomes not only tame but also really affectionate. Certainly, other small birds too become completely tame in the sense that they do not fear their keeper and will sit on his head or shoulder and take titbits from his hands. With a robin this can be achieved in a very short time. However, if one has learned to look deeper into the animal mind and has ceased to project one's own feelings into the creature in the belief that it must love its keeper because he loves it, then one finally sees in the dark, mysterious eyes of the robin only the one somewhat shallow question, "For goodness' sake when am I going to get that mealworm?" The siskin, on the other hand, is a seed-eater that eats the whole day long, is never really hungry and in whose span of interest therefore the ingestion of food plays a smaller role than in that of the insect-hunting bird. The mealworm in the hand of the keeper is a much stronger bait for the robin than is the hemp-seed for the siskin. Therefore the newly caught robin will eat much sooner from the hand than the siskin under like circumstances. Thus the robin can be trained, in a surprisingly short time, to approach its keeper voluntarily; the siskin will only do so after several months, but, once it has taken this step, it approaches him for his company's sake and not in the expectancy of food. Such a "companionable tameness" is much more endearing to our human mentality than the highly material cupboard-love of the robin. As a social animal the siskin can contrive a personal attachment to its keeper of which the robin is incapable. Of course, there are many other social animals that transfer their social impulse to mankind and, when reared young enough, enter into a close social contact with human beings. The starling, the bullfinch and the hawfinch become delightfully affectionate, and the large corvines, parrots, geese and cranes vie even with dogs in this respect.

But all these birds must be taken quite young from the nest if they are to be made into tame and friendly household pets. Why the siskin is an exception to this rule and can find social contact with man even when captured in maturity, nobody knows.

Of the many objects which amply repay the trouble of their care, I mentioned the aquarium, the bullfinch, the starling and the siskin first because they are so easy to keep. Of course, there are dozens of easily obtainable species of animals that are equally easy to maintain, and still more species which are only a little more demanding, and I should strongly advise the beginner to confine himself to this type of animal and to refrain from taking into his charge any really exacting beings.

"Easy to keep" is a quality which must be differentiated sharply from the conceptions "hardy" or "resistant". By keeping a living thing in the scientific sense we understand the attempt to let its whole life cycle be performed before our eyes within the narrower or wider confines of captivity. Nevertheless, those animals are usually deceptively termed easy to keep which, in reality, are merely resistant and, to put it crudely, take a long time to die. The classical example of this type of animal is the Greek tortoise. Even under the inadequate treatment of the average ignorant owner, this poor beast takes three, four or even five years until it is really, thoroughly and irrevocably dead, but, strictly speaking, it starts on the downward path from the first day of its captivity. To keep tortoises so that they grow, thrive and multiply, they must be offered conditions of life which, in a town flat, cannot be achieved. In our own climate nobody, to my knowledge, has truly succeeded in breeding these animals.

When I enter the room of a plant-lover and see that all his plants are growing and flourishing in their present habitat, then I know I have found a soul-mate. I cannot endure having in my room plants that are dying, however slowly they are doing it. The stoutly growing gum-tree, the lusty philodendron, the modest aspidistra which can even thrive in boarding-houses, all warm my heart by their undeniable healthiness, whereas the loveliest rhododendron or cyclamen plant, which is not really growing but slowly deteriorating, brings the breath of putrefaction into my room. For, as Shakespeare says, "If that flower with base infection meet, the basest weed outbraves his dignity". I am also no friend of cut flowers, but their swift death by decapitation disturbs me less than the prolonged sickness of plants deprived of their natural requirements.

Concerning plants, this way of thinking may seem exaggerated, but in the case of animals almost anyone will agree with me. The death of an animal will awake sympathy even in a person less susceptible to suffering. So it is imperative to take on only such animals as, under the conditions that can be offered, really live instead of just dying slowly. Most of the disappointments that later discourage people from keeping animals are attributable to the unfortunate choice of the first one with which they made the attempt. The dead goldfinch lying on the floor of its cage makes a far more lasting impression than the wilting flower in its pot, and the owner, plagued with remorse, swears never again to keep a bird. Had he kept, instead of a goldfinch, a starling or a siskin, he would probably have kept it for fifteen years.

There are few birds that are so often "killed by kindness" by ignorant bird-lovers as goldfinches. They need large amounts of oil-containing seeds, and I myself would hardly undertake the care of a newly caught goldfinch had I not the necessary quantity of thistle- and poppy-seeds at my disposal. The only possible substitute for these is crushed hemp, with emphasis on the "crushed", because the goldfinch is unable to crack open whole hemp-seeds with his rather feeble beak. There are some conscientious dealers of my acquaintance who actually submit their customers to quite a serious examination before they will entrust them with a bird of one of the more exacting species—a most commendable procedure.

Another sound, if seemingly cheap, piece of advice: keep your hands off sick animals. Catch or buy only a healthy bird, take it out of the nest or get it from somebody who understands. If you want to keep an animal for any length of time, do not take on any weaklings and foundlings that are brought to you. The young bird which has fallen from its nest, the roe-deer kid which has strayed from its mother, and all other animals which have fallen by chance into human hands usually bear the seeds of death, or at least they are so weakened that only an owner with veterinary experience can hope to save them. As a general rule, let the procuring of your pet cost you some trouble or money, or both, and it will bring you in interest at the rate of a hundred per cent. When you have made up your mind what you really want, insist on it. But if you are offered a really tame animal, particularly of a social kind, that is, an animal that has obviously been reared by hand from infancy, or has been settled in captivity for a very long time, then seize the opportunity, even if it costs four or five times as much as a timid wild thing of the same species.

An important factor, which busy city workers should take into consideration before buying a pet, is the time-table—

that is, their own and that of the animal. If one leaves one's home for work at daybreak and only returns at dusk, and is accustomed to spend the week-end out of doors, away from home, one will derive little pleasure from a song-bird. The consciousness that, before leaving home, one has cared well for the bird so that it is probably now singing merrily, ensures but meagre satisfaction. If, however, you have managed to acquire, with due consideration for your way of living, a pair of the charming dwarf owls, a tame little owl, some small nocturnal mammal or some other animal which is just beginning its daily round as you come home from work, you will always have something to brighten your leisure hours. Small mammals rarely receive, from animal-lovers, the notice which they deserve. It is true that the more interesting species are rather difficult to obtain. Apart from domesticated house-mice and rats, the equally domesticated and therefore somewhat uninteresting guinea-pig is about the only small mammal which dealers regularly have for sale. In recent times, a new species of rodent has been widely bred and has now appeared in the pet shops. This animal is the golden hamster, and I can thoroughly recommend it to anyone who wants something to while away those lazy evening hours when the brain is too tired for higher intellectual pursuits. Even as I write these lines, a sextet of irresistibly funny three-weeks-old golden hamster babies are performing the drollest wrestling match in which the mouse-sized, cuddly, fat fellows roll over and over and, with loud squeaks and feigned savage bites, chase each other in wild hops round their cage. I know of no other rodent that plays in such an intelligent way, quite like dogs and cats, as the golden hamster. It is cheering to have beside you in the room someone who is so joyfully abandoned and can express it with such quaint gracefulness as one of these little fellows.

I think the golden hamster was created expressly for the sake of the poor animal-lover in the city. It combines all the qualities that are pleasant in a house-pet and is nearly free from those that are undesirable. A tame golden hamster never bites, or at least it does so just as rarely as a guinea-pig or a rabbit; the mothers of very small hamster babies must certainly be handled carefully but, there again, only in the immediate precincts of their brood; a yard away from the nest they can be touched with impunity. How pleasant would the squirrel be as a room-mate if he did not climb up everything and mark all gnawable objects with traces of his teeth! The golden hamster hardly ever climbs, and gnaws so little that he can be allowed to run freely about the room where he will do no appreciable damage. Besides this, this animal is externally the neatest little chap, with his fat head, his big eyes, peering so cannily into the world that they give the impression that he is much cleverer than he really is, and the gaily coloured markings of his gold, black and white coat. Then his movements are so comical that he is ever and again the source of friendly laughter when he comes hurrying, as though pushed along, on his little short legs, or when he suddenly stands upright, like a tiny pillar driven into the floor, and, with stiffly pricked ears and bulging eyes, appears to be on the look-out for some imaginary danger.

On the table in the middle of my room, near the desk, stands the nucleus of my golden hamster stud, a simple little terrarium out of which, with the regularity of the calendar, the litters of young hamsters move, as they grow up, into the roomy boxes that will soon leave no more space in my study. In this terrarium lives the brood mother, with her latest litter. Blasé lovers of rare and delicate animals may deride the fact that I am so much affected by so cheap an animal which every five-year-old child can tend. But, to the student of animal behaviour, it is of no consequence how costly or how difficult to keep an animal may be. He is, or ought to be, entirely free from that ambition, common to so many bird- and fish-lovers, to keep just those perishable species that are most difficult to maintain. His interest in an object is determined by the question of how much can be observed from it, and in this respect the modest golden hamster surpasses many expensive and exacting species. So it happens that my eye rests more often on the little terrarium with golden hamsters than on the aviary which stands beyond it and which contains the most rare and valuable item of my livestock collection: a pair of bearded tits sitting on three eggs.

If I want to, I can keep exacting and delicate animals in such a way that their whole life-cycle is enacted in my study before my eyes, and only he who has succeeded in breeding bearded tits in an indoor aviary or has achieved something equally difficult, can afford to smile over my simple golden hamsters and the great delight which I take in them: but he, presumably, will know better than to do so!

Of course, the past master of animal-keeping may be tempted to try his hand with a particularly tricky species for the sheer love of surmounting difficulties, and, for him, such an attempt may be of value as an exercise; but the beginner must stop to consider that in his case a similar undertaking may easily result in sheer cruelty to animals. The endeavour to keep a very exacting species of animal is only justified by its scientific value; when carried out for a mere fancy it becomes ethically dubious. Even the most experienced animal-keeper should consider, before he undertakes the care of a sensitive organism, that not only the written but also the much more stringent unwritten law demands that captive animals must lack nothing that is necessary to their bodily and mental welfare. In the first enthusiasm over the charm and beauty of a new species, we are often too ready to shoulder this serious responsibility. The enthusiasm fades but the responsibility remains and, before we are aware of it, we are loaded with a burden of which we are not so easily relieved. In the little marble paved pool, which reflects a graceful statue in the corner of our glass veranda, I once kept, for more than a year, two dabchicks, minute diving birds, most interesting in their behaviour and charming to behold. These highly specialized divers cannot stand upright on dry land and walk clumsily, step by step. Normally, they hardly ever leave the water, except to climb on to their floating nest. For this reason they were perfectly contented with their little pool and, once settled and tame, they stayed there of their own free will and without the need of a fence. They were indeed a bewitching piece of interior decoration. Unfortunately these most charming of indoor water-birds possess the awkward property that they will eat only live fish not longer than two inches and not shorter than one inch in size. The few mealworms and the odd bits of greenstuff which they eat in addition to their staple diet are insufficient to ward off hunger for even half a day, should there be a dearth of fish. In spite of the large fish containers with their continuous stream of fresh water which I kept for my charges in the cellar, and although the financial side of the question at that time was no object, the continual worry of food organization was nerve-racking. More than once, in the winter of that year, I rushed, in desperation, from one pet shop to another, or, in equal desperation, hacked open the ice in every pool of the near-lying Danube backwaters that gave promise of small fish, merely to tide over fishless days which for my dabchicks would have spelt death. Although I could not make the decision to part with these pocket swans, I sighed with relief in the midst of my sorrow when, one fine summer's day, the pair found its way out through the open window.

One of the most trying things in a room is the bird which flutters through shyness. You have got a chaffinch, he is lovely and sings well. Since you wish to see as well as hear him, you remove the linen cover in which the previous owner, a knowing finch-keeper, has draped the cage. The bird takes no notice and sings as before—but only as long as you do not move. You dare move only very slowly and carefully, otherwise the bird hurls itself wildly against the cage bars till you fear for its scalp and feathers. Now, you think, he will settle down and become tame, but here you are mistaken. I have known, as yet, only very few chaffinches that became accustomed to people moving about unconcernedly in their vicinity. But do you know what it entails to have to avoid, in your own room, every hasty movement for weeks on end? Do you realize what it means when you dare not even shift a chair, in case the stupid bird again knocks off its freshly sprouting head feathers? At your slightest movement, you squint towards the chaffinch cage, in fear and trembling that the infernal fluttering will start up all over again.

Many migratory birds flutter at night, during migration time. Even if the cage has the usual soft roofing and the bird can therefore do itself no serious damage, this nightly fluttering is a most disturbing business not only for the bird, but also for the person who sleeps in the same room. It is not directly due to the migratory urge that the bird storms the bars of its cage, but it is merely awake, cannot sleep, and the urge for movement forces it ever and again to fly off its perch; since it sees nothing in the dark, it knocks blindly against the bars. The only remedy for this nocturnal fluttering is to install a tiny electric bulb in the cage. It need glow but dimly, just enough for the bird to see the bars and its perches. Only since I discovered this method are my nightly peace and my pleasure in our warblers guaranteed.

I cannot warn the would-be bird fancier enough against under-estimating the shrillness of a bird's song which, outside, sounds sweet and mellow. When a male blackbird starts singing lustily in a room, the window-panes actually vibrate and the cups on the tea-table begin to dance lightly. The songs of the warbler species and of most finches are not too loud for indoors, except possibly that of the chaffinch which may become somewhat irritating by the eternal repetition of its trilling strophe. Altogether, birds which possess a single, never varying strophe should be meticulously avoided by nervy people. It is almost inconceivable that there should be people who not only bear with the common quail but indeed keep him specially for his "pick-per-wick". Imagine three pages of this book inscribed solely with these syllables and you have a good imitation of the quail's song! Charming as it may sound in the open air, in a room it has on me the same effect as a cracked gramophone record where the needle always gets stuck in the same place.

Most fraying of all to the nerves is animal suffering. So for this reason, apart from all higher ethical ones, it is urgently recommended to procure only animals of such species as can easily be kept in good health. A tuberculous parrot brings an atmosphere into the house like that of a dying member of the family. Should an animal, in spite of all due caution, become incurably sick, then do not hesitate to accord to it that act of mercy which the doctor, in an analogous case, must deny to his human patient.

The ability to suffer is, in all living creatures, in direct relation to the extent of their development; this applies, above all, to mental suffering. One of the more stupid animals, such as a nightingale or a small rodent, suffers proportionately much less from close confinement than a raven, a parrot or a mongoose, to say nothing of a lemur or a monkey. To treat one of these clever animals really humanely, one must let it loose from time to time. Such occasional leave from the cage as opposed to permanent confinement seems, at first sight, to imply little essential improvement in the life of the animal. Nevertheless, it makes an inestimable difference to the psychological well-being of the animal. As against permanent imprisonment, it makes exactly the same difference as exists between the life of a continually "tied" human worker and that of a convict!

Let loose? But do not the wild things run or fly away immediately? Those clever animals that suffer mentally under permanent cage life are the least likely to do so. All animals except the very lowest are creatures of habit and wish at all costs to maintain their accustomed mode of life. It is for this reason that every animal suddenly let loose after a long confinement would return to its cage if it could find its way into it. Most of the small cage birds are too stupid for this. Only a few small passerines, such as the house sparrow and the sand martin, possess enough "spatial intelligence" to find their way through the windows and doors of a house. These are the only small birds which may occasionally be allowed the privilege of free flight. One must, however, bear in mind that such tame free-flying small birds are beset by particular dangers which, owing to their trustfulness, are much greater than those which threaten the wild-living fellow-members of their species.

The notion, therefore, that a really tame mongoose, fox or monkey, once let loose, must certainly attempt to regain its "precious freedom" for good and all, implies a false anthropomorphization of the animal's motive. It does not want to get away, it only wants to be let out of the cage. It is no problem to prevent the tame raven, mongoose or monkey from running away; the difficulty is to prevent the animal from disturbing your daily work or Sunday evening peace. I have many years of practice at working in the presence of lively animals and still livelier children, but it annoys me when a raven tries to carry off the pages of my

manuscript; when a starling, with the propeller wind of his wings, blows all the papers off my desk, or when a monkey, behind my back, experiments with something breakable so that I must be prepared, every minute, for a violent crash.

When I sit down at my desk to write, every member of my Noah's Ark must return to its cage. Those intelligent beings that set value on being released from their cage can be so well trained that they will go back again on command (all except the mongoose, who will not do so at any price!).

The dreaded command, once given, is followed by regret on the part of the giver because the animal which crawls so quietly and obediently into its cage tempts one to revoke the order, and this, from an educational point of view, would be most detrimental. But the poor creature, squatting, bored to death in its cage, frays the nerves almost more than it did a few minutes ago when it was free. It is just the same when one permits one's little daughter to remain in one's study but strictly forbids her to speak, or in any other way to disturb one. The inward conflict between good behaviour and the pressing desire to ask a question which is reflected dramatically on the little face, is amongst the sweetest things a little daughter can offer. But it disturbs one's work more than a whole horde of starlings, ravens and monkeys.

My old Alsatian bitch Tito had a special knack of making me suffer in this way. She belonged to that exaggeratedly faithful type of dog which has absolutely no private life of its own but can only exist in and beside its master. She would remain lying at my feet, even if I sat for hours and hours at my desk, and she was far too tactful to whine or to call attention to herself by the slightest sign. She just looked at me. And this gaze of the amber-yellow eyes in which was written the question "Are you ever going to take me out?" was like the voice of conscience and easily penetrated the thickest walls. When I had banished her from the room, I knew, nevertheless, that she was now lying before the front door and that the gaze of those amber-yellow eyes was now unwaveringly fixed on the door-knob.

As I read through this chapter, particularly the last pages, I begin to fear that I may have laid too much stress on the negative side of animal-keeping and have dissuaded you altogether from getting a pet. Do not misunderstand me. If I emphasized so strongly which animals you should not keep, I only did so for fear that disappointment and nerve-racking experiences with your first charge would destroy and spoil forever for you the loveliest and most worthwhile and instructive of all hobbies. For I take very seriously the task of awakening, in as many people as possible, a deeper understanding of the awe-inspiring wonder of Nature, and I am fanatically eager to gain proselytes. And if someone who has patiently read this book as far as this, has allowed himself to be inveigled into setting up an aquarium or buying a pair of golden hamsters, then I have probably won a true adherent to the good cause.

8. THE LANGUAGE OF ANIMALS

Learned of every bird its language,
Learned their names and all their secrets,
Talked with them whene'er he met them.

LONGFELLOW

ANIMALS do not possess a language in the true sense of the word. In the higher vertebrates, as also in insects, particularly in the socially living species of both great groups, every individual has a certain number of innate movements and sounds for expressing feelings. It has also innate ways of reacting to these signals whenever it sees or hears them in a fellow-member of the species. The highly social species of birds, such as the jackdaw or the greylag goose, have a complicated code of such signals which are uttered and understood by every bird without any previous experience. The perfect co-ordination of social behaviour which is brought about by these actions and reactions conveys to the human observer the impression that the birds are talking and understanding a language of their own. Of course, this purely innate signal code of an animal species differs fundamentally from human language, every word of which must be learned laboriously by the human child. Moreover, being a genetically fixed character of the species—just as much as any bodily character—this so-called language is, for every individual animal species, ubiquitous in its distribution. Obvious though this fact may seem, it was, nevertheless, with something akin to naïve surprise that I heard the jackdaws in northern Russia "talk" exactly the same familiar "dialect" as my birds at home in Altenberg. The superficial similarity between these animal utterances and human languages diminishes further as it becomes gradually clear to the observer that the animal, in all these sounds and movements expressing its emotions, has in no way the conscious intention of influencing a fellow-member of its species. This is proved by the fact that even geese or jackdaws reared and kept singly make all these signals as soon as the corresponding mood overtakes them. Under these circumstances the automatic and even mechanical character of these signals becomes strikingly apparent and reveals them as entirely different from human words.

In human behaviour, too, there are mimetic signs which automatically transmit a certain mood and which escape one, without or even contrary to one's intention of thereby influencing anybody else: the commonest example of this is yawning. Now the mimetic sign by which the yawning mood manifests itself is an easily perceived optical and acoustical stimulus whose effect is, therefore, not particularly surprising. But, in general, such crude and patent signals are not always necessary in order to transmit a mood. On the contrary, it is characteristic of this particular effect that it is often brought about by diminutive sign stimuli which are hardly perceptible by conscious observation. The mysterious apparatus for transmitting and receiving the sign stimuli which convey moods is age-old, far older than mankind itself. In our own case, it has doubtless degenerated as our word-language developed. Man has no need of minute intention-displaying movements to announce his momentary mood: he can say it in words. But jackdaws or dogs are obliged to "read in each other's eyes" what they are about to do in the next moment. For this reason, in higher and social animals, the transmitting as well as the receiving apparatus of "mood-convection" is much better developed and more highly specialized than in us humans. All expressions of animal emotions—for instance, the "Kia" and "Kiaw" note of the jackdaw—are therefore not comparable to our spoken language, but only to those expressions such as yawning, wrinkling the brow and smiling, which are expressed unconsciously as innate actions and also understood by a corresponding inborn mechanism. The "words" of the various animal "languages" are merely interjections.

Though man may also have numerous gradations of unconscious mimicry, no George Robey or Emil Jannings would be able, in this sense, to convey by mere miming, as the greylag goose can, whether he was going to walk or fly, or to indicate whether he wanted to go home or to venture farther afield, as a jackdaw can do quite easily. Just as the transmitting apparatus of animals is considerably more efficient than that of man, so also is their receiving apparatus. This is not only capable of distinguishing a large number of signals, but, to preserve the above simile, it responds to much slighter transmissions than does our own. It is incredible what minimal signs, completely imperceptible to man, animals will receive and interpret rightly. Should one member of a jackdaw flock that is seeking for food on the ground fly upwards merely to seat itself on the nearest apple-tree and preen its feathers, then none of the others will cast so much as a glance in its direction; but, if the bird takes to wing with intent to cover a longer distance, then it will be joined, according to its authority as a member of the flock, by its spouse or also a larger group of jackdaws, in spite of the fact that it did not emit a single "Kia".

In this case, a man well versed in the ways and manners of jackdaws might also, by observing the minutest intention-displaying movements of the bird, be able to predict—if with less accuracy than a fellow-jackdaw—how far that particular bird was going to fly. There are instances in which a good observer can equal and even surpass an animal in its faculty of "understanding" and anticipating the intentions of its fellow, but in other cases he cannot hope to emulate it. The dog's "receiving set" far surpasses our own analogous apparatus. Everybody who understands dogs knows with what almost uncanny certitude a faithful dog recognizes in its master whether the latter is leaving the room for some reason uninteresting to his pet, or whether the longed-for daily walk is pending. Many dogs achieve even more in this respect. My Alsatian Tito, the great-great-great-great-great-grandmother of the dog I now possess, knew, by "telepathy", exactly which people got on my nerves, and when. Nothing could prevent her from biting, gently but surely, all such

people on their posteriors. It was particularly dangerous for authoritative old gentlemen to adopt towards me, in discussion, the well-known "you are, of course, too young" attitude. No sooner had the stranger thus expostulated than his hand felt anxiously for the place in which Tito had punctiliously chastised him. I could never understand how it was that this reaction functioned just as reliably when the dog was lying under the table and was therefore precluded from seeing the faces and gestures of the people round it: how did she know who I was speaking to or arguing with?

This fine canine understanding of the prevailing mood of

a master is not really telepathy. Many animals are capable of perceiving the smallest movements, withheld from the human eye. And a dog, whose whole powers of concentration are bent on serving his master and who literally "hangs on his every word", makes use of this faculty to the utmost. Horses too have achieved considerable feats in this field. So it will not be out of place to speak here of the tricks which have brought some measure of renown to certain animals.

There have been "thinking" horses which could work out square roots, and a wonder-dog Rolf, an Airedale terrier, which went so far as to dictate its last will and testament to its mistress. All these "counting", "talking" and "thinking" animals "speak" by knocking or barking sounds, whose meaning is laid down after the fashion of a morse code. At first sight their performances are really astounding. You are invited to set the examination yourself and you are put opposite the horse, terrier or whatever animal it is. You ask, how much is twice two; the terrier scrutinizes you intently and barks four times. In a horse, the feat seems still more prodigious for he does not even look at you. In dogs, who watch the examiner closely, it is obvious that their attention is concentrated upon the latter and not by any means on the problem itself. But the horse has no need to turn his eyes towards the examiner since, even in a direction in which the animal is not directly focusing, it can see, by indirect vision, the minutest movement. And it is you yourself who betray, involuntarily to the "thinking" animal, the right solution. Should one not know the right answer oneself, the poor animal would knock or bark on desperately, waiting in vain for the sign which would tell him to stop. As a rule, this sign is forthcoming, since few people are capable, even with the utmost self-control, of withholding an unconscious and involuntary signal. That it is the human being who finds the solution and communicates it was once proved by one of my colleagues in the case of a dachshund which had become quite famous and which belonged to an elderly spinster. The method was perfidious: it consisted in suggesting a wrong solution of all the problems not to the "counting" dog but to his mistress. To this end, my friend made cards on one side of which a simple problem was printed in fat letters. The cards, however, unknown to the dog's owner, were constructed of several layers of transparent paper on the last of which another problem was inscribed in such a manner as to be visible from behind, when the front side was presented to the animal. The unsuspecting lady, seeing, in looking-glass writing, what she imagined to be the problem to be solved, transmitted involuntarily to the dog a solution which did not correspond to that of the problem on the front of the card, and was intensely surprised when, for the first time in her experience, her pet continued to give wrong answers. Before ending the séance, my friend adopted different tactics and presented mistress and dog with a problem which, for a change, the dog could answer and the lady could not: he put before the animal a rag impregnated with the smell of a bitch in season. The dog grew excited, wagged his tail and whined—he knew what he was smelling and a really knowledgeable dog-owner might have known, too, from observing his behaviour. Not so the old lady. When the dog was asked what the rag smelled of, he promptly morsed *her* answer: "Cheese"!

The enormous sensitivity of many animals to certain minute movements of expression, as, for example, the above-described capacity of the dog to perceive the friendly or hostile feelings which his master harbours for another person, is a wonderful thing. It is therefore not surprising that the naïve observer, seeking to assign to the animal human qualities, may believe that a being which can guess even such inward unspoken thoughts must, still more, understand every word that the beloved master utters; now an intelligent dog does understand a considerable number of words, but, on the other hand, it must not be forgotten that the ability to understand the minutest expressional movements is thus acute in animals for the very reason that they lack true speech.

As I have already explained, all the innate expressions of emotion, such as the whole complicated "signal code" of the jackdaw, are far removed from human language. When your dog nuzzles you, whines, runs to the door and scratches it, or puts his paws on the wash-basin under the tap, and looks at you imploringly, he does something that comes far nearer to human speech than anything that a jackdaw or goose can ever "say", no matter how clearly "intelligible" and appropriate to the occasion the finely differentiated expressional sounds of these birds may appear. The dog wants to make you open the door or turn on the tap, and what he does has the specific and purposeful motive of influencing you in a certain direction. He would never perform these movements if you were not present. But the jackdaw or goose merely gives unconscious expression to its inward mood, and the "Kia" or "Kiaw" or the warning sound escapes the bird involuntarily; when in a certain mood, it must utter the corresponding sound, whether or not there is anybody there to hear it.

The intelligible actions of the dog described above are not innate but are individually learned and governed by true insight. Every individual dog has different methods of making himself understood by his master and will adapt his behaviour according to the situation. My bitch Stasie, the great-grandmother of the dog I now possess, having once eaten something which disagreed with her, wanted to go out during the night. I was at that time overworked, and slept very soundly, so that she did not succeed in waking me and indicating her requirements by her usual signs; to her whining and nosing I had evidently only responded by burying myself still deeper in my pillows. This desperate situation finally induced her to forget her normal obedience and to do a thing which was strictly forbidden her: she jumped on my

bed and then proceeded literally to dig me out of the blankets and roll me on the floor. Such an adaptability to present needs is totally lacking in the "vocabulary" of birds: they never roll you out of bed.

Parrots and large corvines are endowed with "speech" in still another sense: they can imitate human words. Here, an association of thought between the sounds and certain experiences is sometimes possible. This imitating is nothing other than the so-called mocking found in many song-birds. Willow warblers, red-backed shrikes and many others are masters of this art. Mocking consists of sounds, learned by imitation, which are not innate and are uttered only while the bird is singing; they have no "meaning" and bear no relation whatsoever to the inborn "vocabulary" of the species. This also applies to starlings, magpies and jackdaws, who not only "mock" birds' voices but also successfully imitate human words. However, the talking of big corvines and parrots is a somewhat different matter. It still bears that character of playfulness and lack of purpose which is also inherent in the mocking of smaller birds and which is loosely akin to the play of more intelligent animals. But a corvine or a parrot will utter its human words independently of song, and it is undeniable that these sounds may occasionally have a definite thought association.

Many grey parrots, as well as others, will say "Good morning" only once a day and at the appropriate time. My friend Professor Otto Koehler possessed an ancient grey parrot which, being addicted to the vice of feather-plucking, was nearly bald. This bird answered to the name of "Geier", which in German means vulture. Geier was certainly no beauty but he redeemed himself by his speaking talents. He said "Good morning" and "Good evening" quite aptly, and when a visitor stood up to depart he said, in a benevolent bass voice, "Na, auf Wiedersehen". But he only said this if the guest really departed. Like a "thinking" dog, he was tuned in to the finest, involuntarily given signs; what these signs were, we never could find out, and we never once succeeded in provoking the retort by staging a departure. But when the visitor really left, no matter how inconspicuously he took his leave, promptly and mockingly came the words "Na, auf Wiedersehen"!

The well-known Berlin ornithologist, Colonel von Lukanus, also possessed a grey parrot which became famous through a feat of memory. Von Lukanus kept, among other birds, a tame hoopoe named "Höpfchen". The parrot, which could talk well, soon mastered this word. Hoopoes unfortunately do not live long in captivity, though grey parrots do; so, after a time, "Höpfchen" went the way of all flesh and the parrot appeared to have forgotten his name—at any rate, he did not say it any more. Nine years later, Colonel von Lukanus acquired another hoopoe, and as the parrot set eyes on him for the first time, he said at once, and then repeatedly, "Höpfchen" . . . "Höpfchen". . . .

In general, these birds are just as slow in learning something new as they are tenacious in remembering what they have once learned. Everyone who has tried to drum a new word into the brain of a starling or a parrot knows with what patience one must apply oneself to this end, and how untiringly one must again and again repeat the word. Nevertheless, such birds can, in exceptional cases, learn to imitate a word which they have heard seldom, perhaps only once. However, this apparently only succeeds when a bird is in an exceptional state of excitement; I myself have seen only two such cases. My brother had, for years, a delightfully tame and lively blue-fronted Amazon parrot named Papagallo, which had an extraordinary talent for speech. As long as he lived with us in Altenberg, Papagallo flew just as freely around as most of my other birds. A talking parrot that flies from tree to tree and at the same time says human words, gives a much more comical effect than one that sits in a cage and does the same thing. When Papagallo, with loud cries of "Where's the Doc?", flew about the district, sometimes in a genuine search for his master, it was positively irresistible.

Still funnier, but also remarkable from a scientific point of view, was the following performance of the bird; Papagallo feared nothing and nobody, with the exception of the chimney-sweep. Birds are very apt to fear things which are up above. And this tendency is associated with the innate dread of the bird of prey swooping down from the heights. So everything that appears against the sky has for them something of the meaning of "bird of prey". As the black man, already sinister in his darkness, stood up on the chimney-stack and became outlined against the sky, Papagallo fell into a panic of fear and flew, loudly screaming, so

far away that we feared he might not come back. Months later, when the chimney-sweep came again, Papagallo was sitting on the weathercock, squabbling with the jackdaws who wanted to sit there too. All at once, I saw him grow long and thin and peer down anxiously into the village street; then he flew up and away, shrieking in raucous tones, again and again, "The chimney-sweep is coming, the chimney-sweep is coming". The next moment, the black man walked through the doorway of the yard!

Unfortunately, I was unable to find out how often Papagallo had seen the chimney-sweep before and how often he had heard the excited cry of our cook which heralded his approach. It was, without a doubt, the voice and intonation of this lady which the bird reproduced. But he had certainly not heard it more than three times at the most and, each time, only once and at an interval of months.

The second case known to me in which a talking bird learned human words after hearing them only once or very few times, concerns a hooded crow. Again it was a whole sentence which thus impressed itself on the bird's memory. "Hansl", as the bird was called, could compete in speaking talent with the most gifted parrot. The crow had been reared by a railwayman in the next village, and it flew about freely and had grown into a well-proportioned, healthy fellow, a good advertisement for the rearing ability of its foster-father. Contrary to popular opinion, crows are not easy to rear and, under the inadequate care which they usually receive, mostly develop into those stunted, half-crippled specimens which are so often seen in captivity. One day some village boys brought me a dirt-encrusted hooded crow whose wings and tail were clipped to small stumps. I was hardly able to recognize, in this pathetic being, the once beautiful Hansl. I bought the bird, as, on principle, I buy all unfortunate animals that the village boys bring me, and this I do partly out of pity and partly because amongst these stray animals there might be one of real interest. And this one certainly was! I rang up Hansl's master who told me that the bird had actually been missing some days and begged me to adopt him till the next moult. So, accordingly, I put the crow in the pheasant pen and gave it concentrated food, so that, in the imminent new moult, it would grow good new wing and tail feathers. At this time, when the bird was, of necessity, a prisoner, I found out that Hansl had a surprising gift of the gab and he gave me the opportunity of hearing plenty! He had, of course, picked up just what you would expect a tame crow to hear that sits on a tree, in the village street, and listens to the "language" of the inhabitants.

I later had the pleasure of seeing this bird recover his full plumage and I freed him as soon as he was fully capable of flight. He returned forthwith to his former master, in Wordern, but continued, a welcome guest, to visit us from time to time. Once he was missing for several weeks, and when he returned I noticed that he had, on one foot, a broken digit which had healed crooked. And this is the whole point of the history of Hansl, the hooded crow. For we know just how he came by this little defect. And from whom do we know it? Believe it or not, Hansl told us himself! When he suddenly reappeared, after his long absence, he knew a new sentence. With the accent of a true street urchin, he said, in lower Austrian dialect, a short sentence which, translated into broad Lancashire, would sound like "Got 'im in t'bloomin' trap!" There was no doubt about the truth of this statement. Just as in the case of Papagallo, a sentence which he had certainly not heard often had stuck in Hansl's memory because he had heard it in a moment of great apprehension, that is immediately after he had been caught. How he got away again Hansl unfortunately did not tell us.

In such cases, the sentimental animal-lover, crediting the creature with human intelligence, will take an oath on it that the bird understands what he says. This, of course, is quite incorrect. Not even the cleverest "talking" birds, which, as we have seen, are certainly capable of connecting their sound-expressions with particular occurrences, learn to make practical use of their powers, to achieve purposefully even the simplest object. Professor Koehler, who can boast of the greatest successes in the science of training animals, and who succeeded in teaching pigeons to count up to six, tried to teach the above-mentioned, talented grey parrot "Geier" to say "Food" when he was hungry and "Water" when he was dry. This attempt did not succeed, nor, so far, has it been achieved by anybody else. The failure in itself is remarkable. Since, as we have seen, the bird is able to connect his sound-utterances with certain occurrences, we should expect him, first of all, to connect them with a purpose; but this, surprisingly, he is unable to do. In all other cases, where an animal learns a new type of behaviour, it does so to achieve some purpose. The most curious types of behaviour may be thus acquired, especially with the object of influencing the human keeper. A most grotesque habit of this kind was learned by a Blumenau's parakeet which belonged to Professor Karl von Frisch. The scientist only let the bird fly freely when he had just watched it have an evacuation of the bowels, so that, for the next ten minutes, his well-kept furniture was not endangered. The parakeet learned very quickly to associate these facts and, as he was passionately fond of leaving his cage, he would force out a minute dropping with all his might, every time Professor von Frisch came near the cage. He even squeezed desperately when it was impossible to produce anything, and really threatened to do himself an injury by the violence of his straining. You just had to let the poor thing out every time you saw him!

Yet the clever "Geier", much cleverer than that little parakeet, could not even learn to say "Food" when he was hungry. The whole complicated apparatus of the bird's syrinx and brain that makes imitation and association of thought possible, appears to have no function in connection with the survival of the species. We ask ourselves vainly what it is there for!

I only know one bird that learned to use a human word when he wanted a particular thing and who thus connected a sound-expression with a purpose, and it is certainly no coincidence that it was a bird of that species which I consider to have the highest mental development of all, namely the raven. Ravens have a certain innate call-note which corresponds to the "Kia" of the jackdaw and has the same meaning—that is, the invitation to others to fly with the bird that utters it. In the raven, this note is a sonorous, deep-throated, and, at the same time, sharply metallic "krackrackrack". Should the bird wish to persuade another of the same species which is sitting on the ground to fly with it, he executes the same kind of movements as described in the chapter on jackdaws: he flies, from behind, close above the other bird and, in passing it, wobbles with his closely folded tail, at the same time emitting a particularly sharp "Krackrackrackrack" which sounds almost like a volley of small explosions.

My raven Roah, so named after the call-note of the young raven, was, even as a mature bird, a close friend of mine and accompanied me, when he had nothing better to do, on long walks and even on skiing tours, or on motor-boat excursions on the Danube. Particularly in his later years he was not only shy of strange people, but also had a strong aversion to places where he had once been frightened or had had any other unpleasant experience. Not only did he hesitate to come down from the air to join me in such places, but he could not bear to see me linger in what he considered to be a dangerous spot. And, just as my old jackdaws tried to make their truant children leave the ground and fly after them, so Roah bore down upon me from behind, and, flying close over my head, he wobbled with his tail and then swept upwards again, at the same time looking backwards over his shoulder to see if I was following. In accompaniment to this sequence of movements—which, to stress the fact again, is entirely innate—Roah, instead of uttering the above-described call-note, said his own name, with human intonation. The most peculiar thing about this was that Roah used the human word for me only. When addressing one of his own species, he employed the normal innate call-note. To suspect that I had unconsciously trained him would obviously be wrong; for this could only have taken place if, by pure chance, I had walked up to Roah at the very moment when he happened to be calling his name, and, at the same time, to be wanting my company. Only if this rather unlikely coincidence of three factors had repeated itself on several occasions could a corresponding association of thought have been formed by the bird, and that certainly was not the case. The old raven must, then, have possessed a sort of insight that "Roah" was my call-note! Solomon was not the only man who could speak to animals, but Roah is, so far as I know, the only animal that has ever spoken a human word to a man, in its right context—even if it was only a very ordinary call-note.

9. THE TAMING OF THE SHREW

Though Nature, red in tooth and claw
With ravine, shrieked against his creed.

TENNYSON, *In Memoriam*

ALL shrews are particularly difficult to keep; this is not because, as we are led proverbially to believe, they are hard to tame, but because the metabolism of these smallest of mammals is so very fast that they will die of hunger within two or three hours if the food supply fails. Since they feed exclusively on small, living animals, mostly insects, and demand, of these, considerably more than their own weight every day, they are most exacting charges. At the time of which I am writing, I had never succeeded in keeping any of the terrestrial shrews alive for any length of time; most of those that I happened to obtain had probably only been caught because they were already ill, and they died almost at once. I had never succeeded in procuring a healthy specimen. Now the order Insectivora is very low in the genealogical hierarchy of mammals and is, therefore, of particular interest to the comparative ethologist. Of the whole group, there was only one representative with whose behaviour I was tolerably familiar, namely the hedgehog, an extremely interesting animal of whose ethology Professor Herter of Berlin has made a very thorough study. Of the behaviour of all other members of the family practically nothing is known. Since they are nocturnal and partly subterranean animals, it is nearly impossible to approach them in field observation, and the difficulty of keeping them in captivity had hitherto precluded their study in the laboratory. So the Insectivores were officially placed on my programme.

First I tried to keep the common mole. It was easy to procure a healthy specimen, caught to order in the nursery gardens of my father-in-law, and I found no difficulty in keeping it alive. Immediately on its arrival it devoured an almost incredible quantity of earthworms which, from the

very first moment, it took from my hand. But as an object of behaviour study it proved most disappointing. Certainly, it was interesting to watch its method of disappearing in the space of a few seconds under the surface of the ground, to study its astoundingly efficient use of its strong, spade-shaped forepaws, and to feel their amazing strength when one held the little beast in one's hand. And again, it was remarkable with what surprising exactitude it located, by smell, from underground, the earthworms which I put on the surface of the soil in its terrarium. But these observations were the only benefits I derived from it. It never became any tamer and it never remained above ground any longer than it took to devour its prey; after this, it sank into the earth as a submarine sinks into the water. I soon grew tired of procuring the immense quantities of living food it required, and after a few weeks I set it free in the garden.

It was years afterwards, on an excursion to that extraordinary lake, the Neusiedlersee, which lies on the Hungarian border of Austria, that I again thought of keeping an insectivore. This large stretch of water, though not thirty miles from Vienna, is an example of the peculiar type of lake found in the open steppes of Eastern Europe and Asia. More than thirty miles long and half as broad, its deepest parts are only about five feet deep and it is much shallower on the average. Nearly half its surface is overgrown with reeds which form an ideal habitat for all kinds of water-birds. Great colonies of white, purple and grey heron and spoonbills live among the reeds and, until a short while ago, glossy ibis were still to be found here. Greylag geese breed here in great numbers, and on the eastern, reedless shore, avocets and many other rare waders can regularly be found. On the occasion of which I am speaking, we, a dozen tired zoologists, under the experienced guidance of my friend Otto Koenig, were wending our way, slowly and painfully, through the forest of reeds. We were walking in single file, Koenig first, I second, with a few students in our wake. We literally left a wake, an inky-black one in pale grey water. In the reed-forests of Lake Neusiedel, you walk knee-deep in slimy black ooze, wonderfully perfumed by sulphuretted-hydrogen-producing bacteria. This mud clings tenaciously and only releases its hold on your foot with a loud, protesting plop at every step.

After a few hours of this kind of wading you discover aching muscles whose very existence you had never suspected. From the knees to the hips you are immersed in the milky, clay-coloured water characteristic of the lake, which, among the reeds, is populated by myriads of extremely hungry leeches conforming to the old pharmaceutical recipe, "Hirudines medicinales maxime affamati". The rest of your person inhabits the upper air, which here consists of clouds of tiny mosquitoes whose bloodthirsty attacks are all the more exasperating because you require both your hands to part the dense reeds in front of you and can only slap your face at intervals. The British ornithologist who may perhaps have envied us some of our rare specimens will perceive that bird-watching on Lake Neusiedel is not, after all, an entirely enviable occupation.

We were thus wending our painful way through the rushes when suddenly Koenig stopped and pointed mutely towards a pond, free from reeds, that stretched in front of us. At first I could only see whitish water, dark blue sky and green reeds, the standard colours of Lake Neusiedel. Then, suddenly, like a cork popping up on to the surface, there appeared, in the middle of the pool, a tiny black animal, hardly bigger than a man's thumb. And for a moment I was in the rare position of a zoologist who sees a specimen and is not able to classify it, in the literal sense of the word: I did not know to which class of vertebrates the object of my gaze belonged. For the first fraction of a second I took it for the young of some diving bird of a species unknown to me. It appeared to have a beak and it swam on the water like a bird, not in it as a mammal. It swam about in narrow curves and circles, very much like a whirligig beetle, creating an extensive wedge-shaped wake, quite out of proportion to the tiny animal's size. Then a second little beast popped up from below, chased the first one with a shrill, bat-like twitter, then both dived and were gone. The whole episode had not lasted five seconds.

I stood open-mouthed, my mind racing. Koenig turned round with a broad grin, calmly detached a leech that was sticking like a leech to his wrist, wiped away the trickle of blood from the wound, slapped his cheek, thereby killing thirty-five mosquitoes, and asked, in the tone of an examiner, "What was that?" I answered as calmly as I could, "Water-shrews", thanking, in my heart, the leech and the mosquitoes for the respite they had given me to collect my thoughts. But my mind was racing on: water-shrews ate fishes and frogs which were easy to procure in any quantity; water-shrews were less subterranean than most other insectivores; they were the very insectivores to keep in captivity. "That's an animal I must catch and keep," I said to my friend. "That is easy," he responded. "There is a nest with young under the floor mat of my tent." I had slept that night in this tent and Koenig had not thought it worthwhile to tell me of the shrews; such things are, to him, as much a matter of course as wild little spotted crakes feeding out of his hand, or as any other wonders of his queer kingdom in the reeds.

On our return to the tent that evening he showed me the nest. It contained eight young which, compared with their mother, who rushed away as we lifted the mat, were of enormous size. They were considerably more than half her length and must each have weighed well between a fourth and a third of their dam: that is to say, the whole litter weighed, at a very modest estimate, twice as much as the old shrew. Yet they were still quite blind and the tips of their teeth were only just visible in their rosy mouths. And two days later when I took them under my care, they were still quite unable to eat even the soft abdomens of grasshoppers, and in spite of evident greed they chewed interminably on a soft piece of frog's meat without succeeding in detaching a morsel from it. On our journey home I fed them on the squeezed-out insides of grasshoppers and finely minced frog's meat, a diet on which they obviously throve. Arrived home in Altenberg, I improved on this diet by preparing a food from the squeezed-out insides of mealworm larvæ, with some finely chopped small, fresh fishes, worked into a sort of gravy with a little milk. They consumed large quantities of this food, and their little nest-box looked quite small in comparison with the big china bowl whose contents they emptied three times a day. All these observations raise the problem of how the female water-shrew succeeds in feeding her gigantic litter. It is absolutely impossible that she should do so on milk alone. Even on a more concentrated diet my young shrews devoured the equivalent of their own weight

daily, and this meant nearly twice the weight of a grown shrew. Yet, at that time of their lives, young shrews could not possibly engulf a frog or a fish brought whole to them by their mother, as my charges indisputably proved. I can only think that the mother feeds her young by regurgitation of chewed food. Even thus, it is little short of miraculous that the adult female should be able to obtain enough meat to sustain herself and her voracious progeny.

When I brought them home my young water-shrews were still blind. They had not suffered from the journey and were as sleek and fat as one could wish. Their black, glossy coats were reminiscent of moles, but the white colour of their underside, as well as the round, streamlined contours of their bodies, reminded me distinctly of penguins, and not, indeed, without justification: both the streamlined form and the light underside are adaptations to a life in the water. Many free-swimming animals, mammals, birds, amphibians and fishes, are silvery-white below in order to be invisible to enemies swimming in the depths. Seen from below, the shining white belly blends perfectly with the reflecting surface film of the water. It is very characteristic of these water animals that the dark dorsal and the white ventral colours do not merge gradually into each other as is the case in "counter-shaded" land animals whose colouring is calculated to make

them invisible by eliminating the contrasting shade on their undersides. As in the killer whale, in dolphins, and in penguins, the white underside of the water-shrew is divided from the dark upper side by a sharp line which runs, often in very decorative curves, along the animal's flank. Curiously enough, this borderline between black and white showed considerable variations in individuals and even on both sides of one animal's body. I welcomed this, since it enabled me to recognize my shrews personally.

Three days after their arrival in Altenberg my eight shrew babies opened their eyes and began, very cautiously, to explore the precincts of their nest-box. It was now time to remove them to an appropriate container, and on this question I expended much hard thinking. The enormous quantity of food they consumed and, consequently, of excrement they produced, made it impossible to keep them in an ordinary aquarium whose water, within a day, would have become a stinking brew. Adequate sanitation was imperative for particular reasons; in ducks, grebes and all water-fowl, the plumage must be kept perfectly dry if the animal is to remain in a state of health, and the same premise may reasonably be expected to hold good of the shrew's fur. Now water which has been polluted soon turns strongly alkaline, and this I knew to be very bad for the plumage of water-birds. It causes saponification of the fat to which the feathers owe their waterproof quality, and the bird becomes thoroughly wet and is unable to stay on the water. I hold the record, as far as I know hitherto unbroken by any other bird-lover, for having kept dabchicks alive and healthy in captivity for nearly two years, and even then they did not die but escaped, and may still be living. My experience with these birds proved the absolute necessity of keeping the water perfectly clean; whenever it became a little dirty I noticed their feathers beginning to get wet, a danger which they anxiously tried to counteract by constantly preening themselves. I had, therefore, to keep these little grebes in crystal-clear water which was changed every day, and I rightly assumed that the same would be necessary for my water-shrews.

I took a large aquarium tank, rather over a yard in length and about two feet wide. At each end of this I placed two little tables, and weighed them down with heavy stones so that they would not float. Then I filled up the tank until the water was level with the tops of the tables. I did not at first push the tables close against the panes of the tank, which was rather narrow, for fear that the shrews might become trapped under water in the blind alley beneath a table and drown there; this precaution, however, subsequently proved unnecessary. The water-shrew, which in its natural state swims great distances under the ice, is quite able to find its way to the open surface in much more difficult situations. The nest-box, which was placed on one of the tables, was equipped with a sliding shutter, so that I could imprison the

shrews whenever the container had to be cleaned. In the morning, at the hour of general cage-cleaning, the shrews were usually at home and asleep, so that the procedure caused them no appreciable disturbance. I will admit that I take great pride in devising, by creative imagination, suitable containers for animals of which nobody, myself included, has had any previous experience, and it was particularly gratifying that the contraption described above proved so satisfactory that I never had to alter even the minutest detail.

When first my baby shrews were liberated in this container they took a very long time to explore the top of the table on which their nest-box was standing. The water's edge seemed to exert a strong attraction; they approached it ever and again, smelled the surface and seemed to feel along it with the long, fine whiskers which surround their pointed snouts like a halo and represent not only their most important organ of touch but the most important of all their sensory organs. Like other aquatic mammals, the water-shrew differs from the terrestrial members of its class in that its nose, the guiding organ of the average mammal, is of no use whatsoever in its under-water hunting. The water-shrew's whiskers are actively mobile like the antennæ of an insect or the fingers of a blind man.

Exactly as mice and many other small rodents would do under similar conditions, the shrews interrupted their careful exploration of their new surroundings every few minutes to dash wildly back into the safe cover of their nest-box. The survival value of this peculiar behaviour is evident: the animal makes sure, from time to time, that it has not lost its way and that it can, at a moment's notice, retreat to the one place it knows to be safe. It was a queer spectacle to see those podgy black figures slowly and carefully whisker their way forward and, in the next second, with lightning speed, dash back to the nest-box. Queerly enough, they did not run straight through the little door, as one would have expected, but in their wild dash for safety they jumped, one and all, first on to the roof of the box and only then, whiskering along its edge, found the opening and slipped in with a half-somersault, their back turned nearly vertically downward. After many repetitions of this manœuvre they were able to find the opening without feeling for it; they "knew" perfectly its whereabouts yet still persisted in the leap on to the roof. They jumped on to it and immediately vaulted in through the door, but they never, as long as they lived, found out that the leap and vault which had become their habit were really quite unnecessary and that they could have run in directly without this extraordinary detour. We shall hear more about this dominance of path-habits in the water-shrew presently.

It was only on the third day, when the shrews had become thoroughly acquainted with the geography of their little rectangular island, that the largest and most enterprising of them ventured into the water. As is so often the case with mammals, birds, reptiles and fishes, it was the largest and most handsomely coloured male which played the role of leader. First he sat on the edge of the water and thrust in the forepart of his body, at the same time frantically paddling with his forelegs but still clinging with his hind ones to the board. Then he slid in, but in the next moment took fright, scampered madly across the surface very much after the manner of a frightened duckling, and jumped out on to the board at the opposite end of the tank. There he sat, excitedly grooming his belly with one hind paw, exactly as coypus and beavers do. Soon he quietened down and sat still for a moment. Then he went to the water's edge a second time, hesitated for a moment, and plunged in; diving immediately, he swam ecstatically about under water, swerving upward and downward again, running quickly along the bottom, and finally jumping out of the water at the same place as he had first entered it.

When I first saw a water-shrew swimming I was most struck by a thing which I ought to have expected but did not: at the moment of diving, the little black-and-white beast appears to be made of silver. Like the plumage of ducks and grebes, but quite unlike the fur of most water-mammals, such as seals, otters, beavers or coypus, the fur of the water-shrew remains absolutely dry under water—that is to say, it retains a thick layer of air while the animal is below the surface. In the other mammals mentioned above, it is only the short, woolly undercoat that remains dry, the superficial hair-tips becoming wet, wherefore the animal looks its natural colour when under water and is superficially wet when it emerges. I was already aware of the peculiar qualities of the waterproof fur of the shrew, and, had I given it a thought, I should have known that it would look, under water, exactly like the air-retaining fur on the underside of a water-beetle or on the abdomen of a water-spider. Nevertheless, the wonderful, transparent silver coat of the shrew was, to me, one of those delicious surprises that nature has in store for her admirers.

Another surprising detail which I only noticed when I saw my shrews in the water was that they have a fringe of stiff, erectile hairs on the outer side of their fifth toes and on the underside of their tails. These form collapsible oars and a collapsible rudder. Folded and inconspicuous as long as the animal is on dry land, they unfold the moment it enters the water and broaden the effective surface of the propelling feet and of the steering tail by a considerable area.

Like penguins, the water-shrews looked rather awkward and ungainly on dry land but were transformed into objects of elegance and grace on entering the water. As long as they walked, their strongly convex underside made them look pot-bellied and reminiscent of an old, overfed dachshund. But under water, the very same protruding belly balanced harmoniously the curve of their back and gave a beautifully symmetrical streamline which, together with their silver coating and the elegance of their movements, made them a sight of entrancing beauty.

When they had all become familiar with the water their container was one of the chief attractions that our research station had to offer to any visiting naturalists or animal-lovers. Unlike all other mammals of their size, the water-shrews were largely diurnal and, except in the early hours of the morning, three or four of them were constantly on the scene. It was exceedingly interesting to watch their movements upon and under the water. Like the whirligig beetle, Gyrinus, they could turn in an extremely small radius without diminishing their speed, a faculty for which the large rudder surface of the tail with its fringe of erectile hairs is evidently essential. They had two different ways of diving, either by taking a little jump as grebes or coots do and working their way down at a steep angle, or by simply lowering their snout under the surface and paddling very fast till they reached "planing speed", thus working their way downward on the principle of the inclined plane—in other words, performing the converse movement of an ascending aeroplane. The water-shrew must expend a large amount of energy in staying down, since the air contained in its fur exerts a strong pull upwards. Unless it is paddling straight downwards—a thing it rarely does—it is forced to maintain a constant minimum speed, keeping its body at a slightly downward angle, in order not to float to the surface. While swimming under water the shrew seems to flatten, broadening its body in a peculiar fashion, in order to present a better planing surface to the water. I never saw my shrews try to cling by their claws to any underwater objects, as the dipper is alleged to do. When they seemed to be running along the bottom, they were really swimming close above it, but perhaps the smooth gravel on the bottom of the tank was unsuitable for holding on to and it did not occur to me then to offer them a rougher surface. They were very playful when in the water and chased one another loudly twittering on the surface, or silently in the depths. Unlike any other mammal, but just like water-birds, they could rest on the surface; this they used to do, rolling partly over and grooming themselves. Once out again, they instantly proceeded to clean their fur—one is almost tempted to say "preen" it, so similar was their behaviour to that of ducks which have just left the water after a long swim.

Most interesting of all was their method of hunting under water. They came swimming along with an erratic course, darting a foot or so forward very swiftly in a straight line, then starting to gyrate in looped turns at reduced speed. While swimming straight and swiftly their whiskers were, as far as I could see, laid flat against their head, but while circling they were erect and bristled out in all directions, as they sought contact with some prey. I have no reason to believe that vision plays any part in the water-shrew's hunting, except perhaps in the activation of its tactile search. My shrews may have noticed visually the presence of the live tadpoles or little fishes which I put in the tank, but in the actual hunting of its prey the animal is exclusively guided by its sense of touch, located in the wide-spreading whiskers on its snout. Certain small free-swimming species of catfish find their prey by exactly the same method. When these fishes swim fast and straight the long feelers on their snout are depressed, but, like the shrew's whiskers, are stiffly spread out when the fish becomes conscious of the proximity of potential prey; like the shrew, the fish then begins to gyrate blindly in order to establish contact with its prey. It may not even be necessary for the water-shrew actually to touch its prey with one of its whiskers. Perhaps, at very close range, the water vibration caused by the movements of a small fish, a tadpole or a water-insect is perceptible by those sensitive tactile organs. It is quite impossible to determine this question by mere observation, for the action is much too quick for the human eye. There is a quick turn and a snap and the shrew is already paddling shorewards with a wriggling creature in its maw.

In relation to its size, the water-shrew is perhaps the most terrible predator of all vertebrate animals, and it can even vie with the invertebrates, including the murderous Dytiscus larva described in the third chapter of this book. It has been reported by A. E. Brehm that water-shrews have killed fish more than sixty times heavier than themselves by biting out their eyes and brain. This happened only when the fish were confined in containers with no room for escape. The same story has been told to me by fishermen on Lake Neusiedel, who could not possibly have heard Brehm's report. I once offered to my shrews a large edible frog. I never did it again, nor could I bear to see out to its end the cruel scene that ensued. One of the shrews encountered the frog in the basin and instantly gave chase, repeatedly seizing hold of the creature's legs; although it was kicked off again it did not cease in its attack, and finally the frog, in desperation, jumped out of the water and on to one of the tables, where several shrews raced to the pursuer's assistance and buried their teeth in the legs and hindquarters of the wretched frog. And now, horribly, they began to eat the frog alive, beginning just where each one of them happened to have hold of it; the poor frog croaked heart-rendingly, as the jaws of the shrews munched audibly in chorus. I need hardly be blamed for bringing this experiment to an abrupt and agitated end and putting the lacerated frog out of its misery. I never offered the shrews large prey again but only such as would be killed at the first bite or two. Nature can be very cruel indeed; it is not out of pity that most of the larger predatory animals kill their prey quickly. The lion has to finish off a big antelope or a buffalo very quickly indeed in order not to get hurt itself, for a beast of prey which has to hunt daily cannot afford to receive even a harmless scratch in effecting a kill; such scratches would soon add up to such an extent as to put the killer out of action. The same reason has forced the python and other large snakes to evolve a quick and really humane method of killing the well-armed mammals that are their natural prey. But where there is no danger of the victim doing damage to the killer, the latter shows no pity whatsoever. The hedgehog, which by virtue of its armour is quite immune to the bite of a snake, regularly proceeds to eat it, beginning at the tail or in the middle of its body, and in the same way the water-shrew treats its innocuous prey. But man should abstain from judging his innocently-cruel fellow-creatures, for even if nature sometimes "shrieks against his creed", what pain does he himself not inflict upon the living creatures that he hunts for pleasure and not for food?

The mental qualities of the water-shrew cannot be rated very high. They were quite tame and fearless of me and never tried to bite when I took them in my hand, nor did they ever try to evade it, but, like little tame rodents, they tried to dig their way out if I held them for too long in the hollow of my closed fist. Even when I took them out of their container and put them on a table or on the floor, they were by no means thrown into a panic but were quite ready to take food out of my hand and even tried actively to creep into it if they felt a longing for cover. When, in such an unwonted environment, they were shown their nest-box, they plainly showed that they knew it by sight and instantly made for it, and even pursued it with upraised heads if I moved the box along above them, just out of their reach. All in all, I really may pride myself that I have tamed the shrew, or at least one member of that family.

In their accustomed surroundings, my shrews proved to be very strict creatures of habit. I have already mentioned the remarkable conservatism with which they persevered in their unpractical way of entering their nest-box by climbing on to its roof and then vaulting, with a half-turn, in through the door. Something more must be said about the unchanging tenacity with which these animals cling to their habits once they have formed them. In the water-shrew the path-habits, in particular, are of a really amazing immutability; I hardly know another instance to which the saying, "As the twig is bent, so the tree is inclined", applies so literally.

In a territory unknown to it, the water-shrew will never run fast except under pressure of extreme fear, and then it will run blindly along, bumping into objects and usually getting caught in a blind alley. But unless the little animal is severely frightened, it moves, in strange surroundings, only step by step, whiskering right and left all the time and following a path that is anything but straight. Its course is determined by a hundred fortuitous factors when it walks that way for the first time. But, after a few repetitions, it is evident that the shrew recognizes the locality in which it finds itself and that it repeats, with the utmost exactitude, the movements which it performed the previous time. At the same time, it is noticeable that the animal moves along much faster whenever it is repeating what it has already learned. When placed on a path which it has already traversed a few times, the shrew starts on its way slowly, carefully whiskering. Suddenly it finds known bearings, and now rushes forward a short distance, repeating exactly every step and turn which it executed on the last occasion. Then, when it comes to a spot where it ceases to know the way by heart, it is reduced to whiskering again and to feeling its way step by step. Soon, another burst of speed follows and the same thing is repeated, bursts of speed alternating with very slow progress. In the beginning of this process of learning their way, the shrews move along at an extremely slow average rate and the little bursts of speed are few and far between. But gradually the little laps of the course which have been "learned by heart" and which can be covered quickly begin to increase in length as well as in number until they fuse and the whole course can be completed in a fast, unbroken rush.

Often, when such a path-habit is almost completely formed, there still remains one particularly difficult place where the shrew always loses its bearings and has to resort to its senses of smell and touch, sniffing and whiskering vigorously to find out where the next reach of its path "joins on". Once the shrew is well settled in its path-habits it is as strictly bound to them as a railway engine to its tracks and as unable to deviate from them by even a few centimetres. If it diverges from its path by so much as an inch, it is forced to stop abruptly, and laboriously regain its bearings. The same behaviour can be caused experimentally by changing some small detail in the customary path of the animal. Any major alteration in the habitual path threw the shrews into complete confusion. One of their paths ran along the wall adjoining the wooden table opposite to that on which the nest-box was situated. This table was weighted with two stones lying close to the panes of the tank, and the shrews, running along the wall, were accustomed to jump on and off the stones which lay right in their path. If I moved the stones out of the runway, placing both together in the middle of the table, the shrews would jump right up into the air in the place where the stone should have been; they came down with a jarring bump, were obviously disconcerted and started whiskering cautiously right and left, just as they behaved in an unknown environment. And then they did a most interesting thing: they went back the way they had come, carefully feeling their way until they had again got their bearings. Then, facing round again, they tried a second time with a rush and jumped and crashed down exactly as they had done a few seconds before. Only then did they seem to realize that the first fall had not been their own fault but was due to a change in the wonted pathway, and now they proceeded to explore the alteration, cautiously sniffing and bewhiskering the place where the stone ought to have been. This method of going back to the start and trying again always reminded me of a small boy who, in reciting a poem, gets stuck and begins again at an earlier verse.

In rats, as in many small mammals, the process of forming a path-habit—for instance in learning a maze—is very similar to that just described; but a rat is far more adaptable in its behaviour and would not dream of trying to jump over a stone which was not there. The preponderance of motor habit over present perception is a most remarkable peculiarity of the water-shrew. One might say that the animal actually disbelieves its senses if they report a change of environment which necessitates a sudden alteration in its motor habits. In a new environment a water-shrew would be perfectly able to see a stone of that size and consequently to avoid it or to run over it in a manner well adapted to the spatial conditions; but once a habit is formed and has become ingrained, it supersedes all better knowledge. I know of no animal that is a slave to its habits in so literal a sense as the water-shrew. For this animal the geometric axiom that a straight line is the shortest distance between two points simply does not hold good. To them, the shortest line is always the accustomed path, and to a certain extent they are justified in adhering to this principle: they run with amazing speed along their pathways and arrive at their destination much sooner than they would if, by whiskering and nosing, they tried to go straight. They will keep to the wonted path, even though it winds in such a way that it crosses and recrosses itself. A rat or mouse would be quick to discover that it was making an unnecessary detour, but the water-shrew is no more able to do so than is a toy train to turn off at right angles at a level crossing. In order to change its route, the water-shrew must change its whole path-habit, and this cannot be done at a moment's notice but gradually, over a long period of time. An unnecessary, loop-shaped detour takes weeks and weeks to become a little shorter, and after months it is not even approximately straight. The biological advantage of such a path-habit is obvious: it compensates the shrew for being nearly blind and enables it to run exceedingly fast without wasting a minute on orientation. On the other hand it may, under unusual circumstances, lead the shrew to destruction. It has been reported, quite plausibly, that water-shrews have broken their necks by jumping into a pond which had been recently drained. In spite of the possibility of such mishaps, it would be shortsighted if one were simply to stigmatize the water-shrew as stupid because it solves the spatial problems of its daily life in quite a different way from man. On the contrary, if one thinks a little more deeply, it is very wonderful that the same result, namely a perfect orientation in space, can be brought about in two so widely divergent ways: by true observation, as we achieve it, or, as the water-shrew does, by learning by heart every possible spatial contingency that may arise in a given territory.

Among themselves, my water-shrews were surprisingly good-natured. Although, in their play, they would often chase each other, twittering with a great show of excitement, I never saw a serious fight between them until an unfortunate accident occurred: one morning I forgot to reopen the little door of the nest-box after cleaning out their tank. When at last I remembered, three hours had elapsed—a very long time for the swift metabolism of such small insectivores. Upon the opening of the door, all the shrews rushed out and made a dash for the food-tray. In their haste to get out, not only did they soil themselves all over but they apparently discharged, in their excitement, some sort of glandular secretion, for a strong, musk-like odour accompanied their exit from the box. Since they appeared to have incurred no damage by their three hours' fasting, I turned away from the box to occupy myself with other things. However, on nearing the container soon afterwards I heard an unusually loud, sharp twittering and, on my hurried approach, found my

eight shrews locked in deadly battle. Two were even then dying and, though I consigned them at once to separate cages, two more died in the course of the day. The real cause of this sudden and terrible battle is hard to ascertain but I cannot help suspecting that the shrews, owing to the sudden change in the usual odour, had failed to recognize each other and had fallen upon each other as they would have done upon strangers. The four survivors quietened down after a certain time and I was able to reunite them in the original container without fear of further mishap.

I kept those four remaining shrews in good health for nearly seven months and would probably have had them much longer if the assistant whom I had engaged to feed them had not forgotten to do so. I had been obliged to go to Vienna and, on my return in the late afternoon, was met by that usually reliable fellow, who turned pale when he saw me, thereupon remembering that he had forgotten to feed the shrews. All four of them were alive but very weak; they ate greedily when we fed them but died none the less within a few hours. In other words, they showed exactly the same symptoms as the shrews which I had formerly tried to keep; this confirmed my opinion that the latter were already dying of hunger when they came into my possession.

To any advanced animal-keeper who is able to set up a large tank, preferably with running water, and who can obtain a sufficient supply of small fish, tadpoles and the like, I can recommend the water-shrew as one of the most gratifying, charming and interesting objects of care. Of course, it is a somewhat exacting charge. It will eat raw chopped heart (the customary substitute for small live prey) only in the absence of something better, and it cannot be fed exclusively on this diet for long periods. Moreover, really clean water is indispensable. But if these clear-cut requirements be fulfilled, the water-shrew will not merely remain alive but will really thrive, nor do I exclude the possibility that it might even breed in captivity.

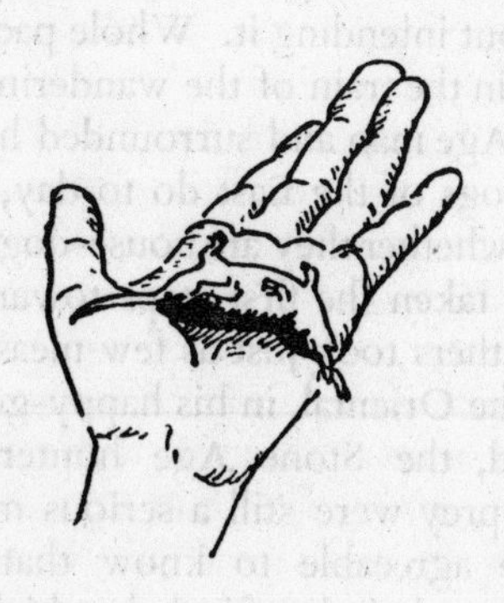

10. THE COVENANT

Four feet trotting behind.

RUDYARD KIPLING

AT the dawn of the later Stone Age there appears, as the first domestic animal, a small semi-domesticated dog, certainly descended from the golden jackal (*Canis aureus*). At this time, in north-west Europe, where skeletons of these dogs have been found, there were probably no more jackals, but there is every reason to believe that the turf dog already lived as a true house-dog and that the lake-dwellers had brought it with them to the shores of the Baltic sea.

But how did Stone Age man come by his dog? Very probably without intending it. Whole packs of jackals must have followed in the train of the wandering, hunting hordes of early Stone Age man and surrounded his settlements just as the pariah dogs of the East do to-day, of whom no one knows exactly whether they are house-dogs run wild, or wild dogs that have taken the first steps towards domestication. And our forefathers took just as few measures against these scavengers as the Oriental, in his happy-go-lucky way, does to-day. Indeed, the Stone Age hunters, for whom the large beasts of prey were still a serious menace, must have found it quite agreeable to know that their camp was watched by a broad circle of jackals which, at the approach of a sabre-toothed tiger or a marauding cave-bear, gave tongue in the wildest tones.

Then, some time or other, to the function of the sentry was added that of a helper in the hunting field. Some time or other, the pack of jackals which used to follow the hunter in the hope of receiving the entrails of his prey took to running before instead of behind the hunter; it began to track game and even to bring it to bay. It is very easy to imagine how these prehistoric dogs developed a new type of interest in the larger game animals. Originally, a jackal would show no interest in the trail of a stag or wild horse, since by himself he could not hope to kill and eat it, but it is not too much to assume that, after having repeatedly received entrails or other refuse from that kind of beast, he might have found inducement to follow a trail which, by its scent, reminded him of a good meal. He might even, by a stroke of canine genius, have "conceived the idea" of calling the hunter's attention to the track. It is remarkable how quickly dogs realize when they can rely on the help of a strong friend. Even my rather cowardly miniature French bulldog would, if accompanied by his friend, a huge Newfoundland, recklessly attack any dog he met. I am not, therefore, crediting the primitive jackal-dogs with too much intelligence when I surmise that, without being consciously trained by man, they learned to track and bring to bay large game animals.

To me it is a strangely appealing and even elevating thought that the age-old covenant between man and dog was "signed" voluntarily and without obligation by each of the contracting parties. All other domestic animals, like some slaves of ancient times, became house-servants only after having served a term of true imprisonment—all, that is, with the exception of the cat; for the cat is not really a domesticated animal and his chief charm lies in the fact that, even to-day, he still walks by himself. Neither the dog nor the cat

is a slave, but only the dog is a friend—granted, a submissive and servile friend. Very gradually, in the course of the centuries, it has become customary, in the "better families" of dogs, to choose, instead of another dog, a man as a leader of their pack. In many cases this appears to have been the chief of a human tribe, and even dogs of to-day, particularly those of strong individual character, tend to consider the "paterfamilias" as their master. In huskies and other primitive breeds, a more complicated and less direct type of submission to man can often be observed. When many of these dogs are kept together one of them stands out as leader, and the others are "faithful" and "respectful" only to him and it is only the leader himself who is, in a true sense, his master's dog; the others are, strictly speaking, the leader's dogs. Reading between the lines, one can tell from Jack London's obviously true-to-life description that in sledge-dog teams this type of relationship is the rule, and it is most probable that it also prevailed among the primitive jackal-dogs of the Stone Age. In modern dogs, however, it is interesting to note that most of them do not seem content with a dog as master and actively seek for a man as leading dog.

One of the most wonderful and puzzling phenomena is the choice of a master by a good dog. Quite suddenly, often within a few days, a bond is formed which is many times stronger than any tie that ever exists between us human beings. Wordsworth calls it:

> . . . that strength of feeling, great
> Above all human estimate.

There is no faith which has never yet been broken, except that of a truly faithful dog. Of all dogs which I have hitherto known, the most faithful are those in whose veins flows, beside that of the golden jackal (*Canis aureus*), a considerable stream of wolf's blood. The northern wolf

(*Canis lupus*) only figures in the ancestry of our present dog breeds through having been crossed with already domesticated Aureus dogs. Contrary to the widespread opinion that the wolf plays an essential role in the ancestry of the larger dog breeds, comparative research in behaviour has revealed the fact that all European dogs, including the largest ones, such as Great Danes and wolf-hounds, are pure Aureus and contain, at the most, a minute amount of wolf's blood. The purest wolf-dogs that exist are certain breeds of Arctic America, particularly the so-called malemuts, huskies, etc. The Esquimaux dogs of Greenland also show but slight traces of Lapland Aureus characters, whereas the Arctic breeds of the Old World such as Lapland dogs, Russian lajkas, samoyedes and chow-chows, certainly have more Aureus in their constitution. Nevertheless the latter breeds derive their character from the Lupus side of their ancestry and they all exhibit the high cheek-bones, the slanting eyes and the slightly upward tilt of the nose which give its specific expression to the face of the wolf. On the other hand, the chow, in particular, bears unquestionably the stamp of his share of Aureus blood in the flaming red of his magnificent coat.

The "sealing of the bond", the final attachment of the dog to one master, is quite enigmatical. It takes place quite suddenly, within a few days, particularly in the case of puppies that come from a breeding kennel. The "susceptible period" for this most important occurrence in the whole of a dog's life is, in Aureus dogs, between eight and eighteen months, and in Lupus dogs round about the sixth month.

The really single-hearted devotion of a dog to its master has two quite different sources. On the one side, it is nothing else than the submissive attachment which every wild dog shows towards his pack leader, and which is transferred, without any considerable alteration in character, by the domestic dog to a human being. To this is added, in the more highly domesticated dogs, quite another form of affection. Many of the characteristics in which domestic animals differ from their wild ancestral form arise by virtue of the fact that properties of body structure and behaviour, which in the wild prototype are only marked by some transient stages of youth, are kept permanently by the domestic form. In dogs, short hair, curly tail, hanging ears, domed skulls and the short muzzle of many domestic breeds are features of this type. In behaviour, one of these juvenile characters which has become permanent in the domestic dog expresses itself in the peculiar form of its attachment.

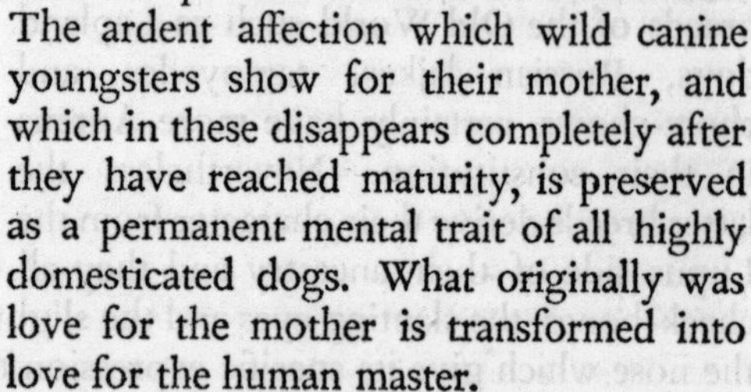

The ardent affection which wild canine youngsters show for their mother, and which in these disappears completely after they have reached maturity, is preserved as a permanent mental trait of all highly domesticated dogs. What originally was love for the mother is transformed into love for the human master.

Thus the pack loyalty, in itself unaltered, but merely transferred to man, and the permanent child-like dependency resulting from domestication are two more or less independent springs of canine affection. One essential difference in the character of Lupus and Aureus dogs is attributable to the fact that these two springs flow with different strength in the two types. In the life of a wolf, the community of the pack plays a vastly more important role than in that of a jackal. While the latter is essentially a solitary hunter and confines himself to a limited territory, the wolf pack roams far and wide through the forests of the North as a sworn and very exclusive band which sticks together through thick and thin and whose members will defend each other to the very death. That the wolves of a pack will devour each other, as is frequently asserted, I have strong reason to doubt, since sledge dogs will not do so at any price, even when at the point of starvation, and this social inhibition has certainly not been instilled into them by man.

The reticent exclusiveness and the mutual defence at any price are properties of the wolf which influence favourably the character of all strongly wolf-blooded dog breeds and distinguish them to their advantage from Aureus dogs, which are mostly "hail-fellow-well-met" with every man and will follow anyone who holds the other end of the lead in his hand. A Lupus dog, on the contrary, who has once sworn allegiance to a certain man, is forever a one-man dog, and no stranger can win from him so much as a single wag of his bushy tail. Nobody who has once possessed the one-man love of a Lupus dog will ever be content with one of pure Aureus blood. Unfortunately this fine characteristic of the Lupus dog has against it various disadvantages which are indeed the immediate results of the one-man loyalty. That a mature Lupus dog can never become *your* dog, is a matter of course. But worse, if he is already yours and you are forced to leave him, the animal becomes literally mentally unbalanced, obeys neither your wife nor children, sinks morally, in his grief, to the level of an ownerless street cur, loses his restraint from killing and, committing misdeed upon misdeed, ravages the surrounding district.

Besides this, a predominantly Lupus-blooded dog is, in spite of his boundless loyalty and affection, never quite sufficiently submissive. He is ready to die for you, but not to obey you: at least, I have never been able to extract implicit obedience from one of these dogs—perhaps a better dog-trainer than I might be more successful. For this

reason, it is seldom that you see, in a town, a chow without a lead and walking close beside his master. If you walk with a Lupus dog in the woods you can never make him stay near you. All he will do is to keep in very loose contact with you and honour you with his companionship only now and again.

Not so the Aureus dog; in him, as a result of his age-old domestication, that infantile affection has persisted which makes him a manageable and tractable companion. Instead of the proud, manly loyalty of the Lupus dog which is far removed from obedience, the Aureus dog will grant you that servitude which, day and night, by the hour and by the minute, awaits your command and even your slightest wish. When you take him for a walk an Aureus dog of a more highly domesticated breed will, without previous training, always run with you, keeping the same radius whether he runs before, behind or beside you and adapting his speed to yours. He is naturally obedient—that is to say, he answers to his name not only when he wishes to and when you cajole him but also because he knows that he *must* come. The harder you shout, the more surely he will come, whereas a Lupus dog, in this case, comes not at all but seeks to appease you from a distance with friendly gestures.

Opposed to these good and congenial properties of the Aureus dog are unfortunately some others which also arise from the permanent infantility of these animals and are less agreeable for an owner. Since young dogs under a certain age are, for members of their own species, "taboo"—that is, they must not under any circumstances be bitten—such big babies are often correspondingly trustful and importunate towards everybody. Like many spoilt human children who call every grown-up "uncle", they pester people and animals alike with overtures to play. If this youthful property persists, to any appreciable extent, in the adult domestic dog, there arises a very unpleasant canine character, or rather the complete lack of such a commodity. The worst part of it lies in the literally "dog-like" submission that these animals, who see in every man an "uncle", show towards anyone who treats them with the least sign of severity; the playful storm of affection is immediately transformed into a cringing state of humility. Everyone is acquainted with this kind of dog which knows no happy medium between perpetual exasperating "jumping up", and fawningly turning upon its back, its paws waving in supplication. You shout, at the risk of offending your hostess, at the infuriating creature that is trampling all over your person and covering you from head to foot with hairs. Thereupon the dog falls beseechingly upon his back. You speak kindly to him, to conciliate your hostess, and—splash—quickly leaping up, the brute has licked you right across the face and now continues unremittingly to bestrew your trousers with hairs.

A dog of this kind, which is everybody's dog, is easily led astray since he trusts every stranger who speaks kindly to him. But a dog that you can get so easily—well, so far as

I am concerned, you can keep him! Even the many alluring and beautifully proportioned breeds of gun-dog, whose "heads are hung with ears that sweep away the morning dew", are uncongenial to my taste in that most of them are ready to follow any man with a gun. Admittedly, their usefulness as gun-dogs is based on this general acceptance of anyone as master—and indeed, were this not so, one could never buy a ready-trained gun-dog or have one's dog schooled by a professional trainer. It is clear that a dog can only be trained by a man who commands his absolute obedience and trust. When you leave your dog with a trainer you therefore imply, from the first, a breach of loyalty. The personal relationship between master and dog must necessarily be severely injured, even if the dog, on his return from the trainer, once more reverts to something of his former attachment to his owner.

Should one do the same thing with a Lupus-blooded dog, he would either learn nothing at all and, through stubborn shyness, if not by sheer aggressive ill-temper, drive his trainer to distraction or, if one sent the dog early enough to the school before his fidelity had found an object on which to rest unshakeably, then, without doubt, the love of the animal would belong for good to his trainer. It is therefore out of the question to buy a Lupus dog as a fully-trained animal. Separated from the master of his choice, the dog would show no signs of ever having been trained. The Lupus dog either accepts one master, unconditionally and for all time, or, if he does not find one or if he loses him, he becomes as independent and self-sufficient as a cat and lives alongside the human being without ever developing any heart-felt connection with him. In this condition most of the North American sledge-dogs find themselves, whose deep qualities of soul are almost never awakened unless a Jack London recognizes and finds access to them. The same holds good for many of our Middle European chows who, for this reason, are despised by many dog-lovers and disliked by most veterinary surgeons. Chows very often "turn cat", in the manner described above, since their first true love often proves unsatisfactory and they are incapable of a second. Chows swear their irrevocable oath of fidelity particularly early. There is almost no Aureus dog, of however true and staunch a character, such as the Airedale terrier or the Alsatian, whose love may not still be won for a completely new master at the age of about a year. But if one wants to be certain of the fidelity of a chow, or any other Lupus-blooded dog, one must rear him oneself from a very early age. Judging from my long experience with chows, one should begin with one of these dogs at the age of four or at the most five months. This is no such great sacrifice as one might expect, for in Lupus dogs the tendency to become house-trained matures much earlier than in those of Aureus breeds. Indeed, the cat-like instinct for cleanliness is amongst the most agreeable qualities of the Lupus species.

Yet my affections do not belong entirely to Lupus dogs, as the reader might conclude from this little canine characterology. No Lupus-blooded dog has so far offered his master such unquestioning obedience as our incomparable Alsatian (an Aureus dog). Admittedly, the noble qualities of the beast of prey possessed by the Lupus dog, his proud aloofness towards strangers, his boundless love for his master, and, at the same time, the reticence with which he demonstrates his really deep affection, are all character traits for which the Aureus dog has no counterpart. But both sets of qualities can be combined. It would, of course, be quite impossible for the dog-breeder to make the predominantly Lupus dog catch up, in one stride, with the Aureus dog which has been domesticated for a few thousand years longer, but there is another way.

Some years ago my wife and I each possessed a dog: I the already-mentioned Alsatian bitch Tito, my wife the little

chow bitch Pygi. Both were true types of their breed, classical representatives of the groups *C. aureus* and *C. lupus*, and they provoked, in their way, some marital strife. My wife disdained me, because Tito used joyfully to greet our family friends; because she would splash through any puddle and then, covered with mud, unconcernedly run through our best rooms; because, on the point of house manners, she left much to be desired if we forgot to let her out—and because of a hundred little sins that a Lupus dog would not commit at any price. Then, said my wife, the dog had no private life, she was just the soulless shadow of her master and it got on one's nerves to see her lying, all day long, beside the desk and, with longing eyes, awaiting the next walk. . . . Shadow! Soulless! Tito this soul of a dog! I replied, in kind, that as far as I was concerned, you could keep a dog that you could not take for a walk, for that was what a dog was for, to follow his master obediently, and Pygi, in spite of her much-praised qualities as a one-man dog, immediately went off hunting—or had my wife ever once returned, accompanied by her dog, from a walk in the forest? You might as well, from the beginning, get a Siamese cat, which was still more aloof, still cleaner, and, above all, what she pretended to be . . a cat. Pygi was not a dog. Nor was my Tito, would come the answer, or, at best, a sentimental figure out of a Victorian novel.

This quarrel, in whose joking tone some earnest was intermixed, found the most natural compromise possible. A son of Tito's, Booby by name, married the chow bitch Pygi. This happened quite against the will of my wife who, naturally enough, wanted to breed pure chows. But here we discovered, as an unexpected hindrance, a new property of Lupus dogs: the monogamous fidelity of the bitch to a certain dog. My wife travelled with her bitch to nearly all the chow dogs in Vienna, in the hope that one at least would find Pygi's favour. In vain—Pygi snapped furiously at all her suitors; she only wanted her Booby and she got him in the end, or rather he got her by reducing a thick wooden door, behind which Pygi was confined, to its primary elements.

And therewith began our chow-alsatian cross-bred stud. The whole credit falls on the true love of Pygi for her enormous and good-natured Booby. The reader should grant me due approval for recording the proceedings faithfully. I might have been tempted to write: "After my intensive analysis of the advantages and disadvantages inherent in the character of Lupus and Aureus dogs, I decided that crossing experiments, with a view to combining the good qualities of both, were called for. These succeeded beyond all anticipation. Whereas, generally, cross-breds inherit the bad properties of both parent breeds, in this case the contrary proved true in a very definite measure . . ." As regards the success, this statement would be quite true, only I must admit that the whole thing took place without any preliminary planning.

At the moment our breed contains very little Alsatian blood, because my wife, during my absence in the war, twice crossed in pure chows; this was inevitable, for without so doing we should have been dependent on inbreeding. As it is, the inheritance of Tito shows itself clearly in a psychological respect, for the dogs are far more affectionate and much easier to train than pure-blooded chows, although, from an external point of view, only a very expert eye can detect the element of Alsatian blood. I intend to develop further this mixed breed, now that it has happily survived the war, and to continue with my plan to evolve a dog of ideal character.

Is it justifiable to create still another breed of dog in addition to the many that are already in existence? I think so. In these days the value of the dog to man is purely a psychological one, except in the few cases where the animal has a utilitarian purpose, as for sportsmen or policemen. The pleasure which I derive from my dog is closely akin to the joy accorded to me by the raven, the greylag goose or other wild animals that enliven my walks through the countryside; it seems like a re-establishment of the immediate bond with that unconscious omniscience that we call nature. The price which man had to pay for his culture and civilization was the severing of this bond which had to be torn to give him his specific freedom of will. But our infinite longing for paradise lost is nothing else than a half-conscious yearning for our ruptured ties. Therefore, I need a dog that is no phantasy of fashion but a living animal, no product of science or triumph of form-breeding art but a natural being with an undistorted soul. And this, unfortunately, is what very few pedigree dogs possess, least of all such breeds as, at some time or other, have become "modern" and have been bred with exclusive consideration for a certain external appearance. So far, every breed of dog that has been exposed to this process has been damaged in mind and soul. I wish to achieve the opposite result: my purpose in breeding dogs is to bring about an ideal combination of the psychological qualities of Lupus and of Aureus dogs. I want to breed a dog which is specially capable of supplying that which poor, civilized, city-pent man is so badly in need of!

Let us admit this and not lie to ourselves that we need the dog as a protection for our house. We *do* need him, but not as a watch-dog. I, at least in dreary foreign towns, have certainly stood in need of my dog's company and I have derived, from the mere fact of his existence, a great sense of inward security, such as one finds in a childhood memory or in the prospect of the scenery of one's own home country, for me the Blue Danube, for you the White Cliffs of Dover. In the almost film-like flitting-by of modern life, a man needs something to tell him, from time to time, that he is still himself, and nothing can give him this assurance in so comforting a manner as the "four feet trotting behind".

11. THE PERENNIAL RETAINERS

Enough, if something from our hands have power
To live, and act, and serve the future hour.

WORDSWORTH, *After-Thought*

IN the chimney the autumn wind sings the song of the elements, and the old firs before my study window wave excitedly with their arms and sing so loudly in chorus that I can hear their sighing melody through the double panes. Suddenly, from above, a dozen black, streamlined projectiles shoot across the piece of clouded sky for which my window forms a frame. Heavily as stones they fall, fall to the tops of the firs where they suddenly sprout wings, become birds and then light feather rags that the storm seizes and whirls out of my line of vision, more rapidly than they were borne into it.

I walk to the window to watch this extraordinary game that the jackdaws are playing with the wind. A game? Yes, indeed, it is a game, in the most literal sense of the word: practised movements, indulged in and enjoyed for their own sake and not for the achievement of a special object. And rest assured, these are not merely inborn, purely instinctive actions, but movements that have been carefully learned. All these feats that the birds are performing, their wonderful exploitation of the wind, their amazingly exact assessment of distances and, above all, their understanding of local wind conditions, their knowledge of all the up-currents, air-pockets and eddies—all this proficiency is no inheritance, but, for each bird, an individually acquired accomplishment.

And look what they do with the wind! At first sight you, poor human being, think that the storm is playing with the birds, like a cat with a mouse, but soon you see, with astonishment, that it is the fury of the elements that here plays the role of the mouse and that the jackdaws are treating the storm exactly as the cat its unfortunate victim. Nearly, but only nearly, do they give the storm its head, let it throw them high, high into the heavens, till they seem to fall upwards, then, with a casual flap of a wing, they turn themselves over, open their pinions for a fraction of a second from below against the wind, and dive—with an acceleration far greater than that of a falling stone—into the depths below. Another tiny jerk of the wing and they return to their normal position and, on close-reefed sails, shoot away with breathless speed into the teeth of the gale, hundreds of yards to the west: this all playfully and without effort, just to spite the stupid wind that tries to drive them towards the east. The sightless monster itself must perform the work of propelling the birds through the air at a rate of well over 80 miles an hour; the jackdaws do nothing to help beyond a few lazy adjustments of their black wings. Sovereign control over the power of the elements, intoxicating triumph of the living organism over the pitiless strength of the inorganic!

* * * * *

Twenty-five years have passed since the first jackdaw flew round the gables of Altenberg and I lost my heart to the bird with the silvery eyes. And, as so frequently happens with the great loves of our lives, I was not conscious of it at the time when I became acquainted with my first jackdaw.

It sat in Rosalia Bongar's pet shop, which still holds for me all the magic of early childhood memories. It sat in a rather dark cage and I bought it for exactly four shillings, not because I intended to use it for scientific observations, but because I suddenly felt a longing to cram that great, yellow-framed red throat with good food. I wished to let it fly as soon as it became independent and this I really did, but with the unexpected consequence that even to-day, after the terrible war, when all my other birds and animals are gone, the jackdaws are still nesting under our roof-tops. No bird or animal has ever rewarded me so handsomely for an act of pity.

Few birds—indeed, few of the higher animals (the colony-building insects come under a different heading)—possess so highly developed a social and family life as the jackdaws. Accordingly, few animal babies are so touchingly helpless and so charmingly dependent on their keeper as young jackdaws. Just as the quills of its primary feathers became hard and ready for flight, my young bird suddenly developed a really child-like affection for my person. It refused to remain by itself for a second, flew after me from one room to another and called in desperation if ever I was forced to leave it alone. I christened it "Jock" after its own call-note, and to this day we preserve the tradition that the first young bird of a new species reared in isolation is christened after the call-note of its kind.

Such a fully fledged young jackdaw, attached to its keeper by all its youthful affection, is one of the most wonderful objects for observation that you can imagine. You can go outside with the bird and, from the nearest viewpoint, watch its flight, its method of feeding, in short all its habits, in perfectly natural surroundings, unhampered by the bars of a cage. I do not think that I have ever learned so much about the essence of animal nature from any of my beasts or birds as from Jock in that summer of 1925.

It must have been owing to my gift of imitating its call that it soon preferred me to any other person. I could take long walks and even bicycle rides with it and it flew after me, faithful as a dog. Although there was no doubt that it knew me personally and preferred me to anybody else, yet it would desert me and fly after some other person if he was walking much faster than me, particularly if he overtook

me. The urge to fly after an object moving away from it is very strong in a young jackdaw and almost takes the form of a reflex action. As soon as he had left me, Jock would notice his error and correct it, coming back to me hurriedly. As he grew older, he learned to repress the impulse to pursue a stranger, even one walking very fast indeed. Yet even then I would often notice his giving a slight start or a movement indicative of flying after the faster traveller.

Jock had to struggle with a still greater mental conflict when one or more hooded crows, common in this district, flew up in front of us. The sight of those beating black wings disappearing rapidly into the distance released in the jackdaw an irresistible urge to pursue which it never, in spite of bitter experience, learned to resist. It used to rush blindly after the crows which repeatedly lured it far away, and it was only by good luck that it did not get lost altogether. Most peculiar was its reaction when the crows alighted: the moment that the magic of those flapping black wings ceased to work, Jock entirely lost interest! Though a flying crow had such an overwhelming attraction for him, a sitting one evidently did not, and as soon as the crows landed he had had enough of them, was seized with loneliness and began to call for me in that strange, complaining tone with which young, lost jackdaws call for their parents. As soon as he heard my answering call he rose and flew towards me with such determination that he frequently drew the crows with him and came flying to my side as the leader of their troop. So blindly would the crows follow him in such cases that they were almost upon me before they noticed me at all. When finally they became conscious of my presence they were struck with terror and darted away in such a panic that Jock—infected by the general consternation—once again flew away after them. When I had learned to recognize this danger I was able to avoid complications by making myself as conspicuous as possible and thus warding off the approaching crows early enough to prevent a panic.

Like the stones of a mosaic, the inherited and acquired elements of a young bird's behaviour are pieced together to produce a perfect pattern. But, in a bird that has been reared by hand, the natural harmony of this design is necessarily somewhat disturbed. All those social actions and reactions whose object is not determined by inheritance, but acquired by individual experience, are apt to become unnaturally deflected. In other words, they are directed towards human beings, instead of fellow-members of the bird's species. As Rudyard Kipling's Mowgli thought of himself as a wolf, so Jock, had he been able to speak, would certainly have called himself a human being. Only the sight of a pair of flapping black wings sounded a hereditary note: "Fly with us". As

long as he was walking, he considered himself a man, but the moment he took to wing, he saw himself as a hooded crow, because these birds were the first to awaken his flock instinct.

When in Kipling's Mowgli love is awakened, this all-powerful urge forces him to leave his wolf brothers and to return to the human family. This poetical assumption is scientifically correct. We have good reason to believe that in human beings—as in most mammals—the potential object of sexual love makes itself evident by characters which speak to the depth of age-old inheritance, and not by signs recognizable by experience—as evidently is the case in many birds. Birds reared in isolation from their kind do not generally know which species they belong to; that is to say, not only their social reactions but also their sexual desires are directed towards those beings with whom they have spent certain impressionable phases of their early youth. Consequently, birds raised singly by hand tend to regard human beings, and human beings only, as potential partners in all reproductive activities. And this is exactly what Jock did.

This phenomenon can be observed regularly in hand-reared male house sparrows, who, for this reason, enjoyed great popularity among the loose-living ladies of Roman society, and whom Catullus has immortalized by his little poem "Passer mortuus est meae puellae". But there is no limit to the queer errors that may arise in this connection. A female barnyard goose which I now possess was the only survivor of a brood of six, of which the remainder all succumbed to avian tuberculosis. Consequently she grew up in the company of chickens and, in spite of the fact that we bought for her, in good time, a beautiful gander, she fell head over heels in love with our handsome Rhode Island cock, inundated him with proposals, jealously prevented him from making love to his hens and remained absolutely insensible to the attentions of the gander. The hero of a similar tragi-comedy was a lovely white peacock of the Schönbrunn Zoo in Vienna. He too was the last survivor of an early-hatched brood which perished in a period of cold weather, and to save him, the keeper put him in the warmest room to be found in the whole Zoo, which at that time, shortly after the first world war, was in the reptile house with the giant tortoises! For the rest of his life this unfortunate bird saw only in those huge reptiles the object of his desire and remained unresponsive to the charms of the prettiest pea-hens. It is typical of this extraordinary state of fixation of sexual desire on a particular and unnatural object that it cannot be reversed.

When Jock reached maturity he fell in love with our housemaid, who just then married and left our service. A few days later Jock discovered her in the next village two miles away, and immediately moved into her cottage, returning only at night to his customary sleeping quarters. In the middle of June, when the mating season of jackdaws was over, he suddenly returned home to us and forthwith adopted one of the fourteen young jackdaws which I had reared that spring. Towards this protégé Jock displayed exactly the same attitude as normal jackdaws show towards their young. The behaviour towards its offspring must, of necessity, be innate in any bird or animal, since its own young are the first with which it becomes acquainted. Did a jackdaw not respond to them with instinctively established, inherited reactions, it would not know how to take care of them and might even tear them to pieces and devour them, like any other living object of the same size.

I must now dispel in the reader an illusion which I myself harboured up to the time when Jock reached sexual maturity; the kind of advances which Jock made to our housemaid slowly but surely divulged the fact that "he" was a female!

She reacted to this young lady exactly as a normal female jackdaw would to her mate. In birds—even in parrots, of which the opposite is often maintained—there is no law of attraction of opposites, by which female animals are drawn towards men and males towards women. Another tame adult male jackdaw fell in love with me and treated me exactly as a female of his own kind. By the hour, this bird tried to make me creep into the nesting cavity of his choice, a few inches in width, and in just the same way a tame male house sparrow tried to entice me into my own waistcoat pocket. The male jackdaw became most importunate in that he continually wanted to feed me with what he considered the choicest delicacies. Remarkably enough, he recognized the human mouth in an anatomically correct way as the orifice of ingestion and he was overjoyed if I opened my lips to him, uttering at the same time an adequate begging note. This must be considered as an act of self-sacrifice on my part, since even I cannot pretend to like the taste of finely minced worm, generously mixed with jackdaw saliva. You will understand that I found it difficult to co-operate with the bird in this manner every few minutes!

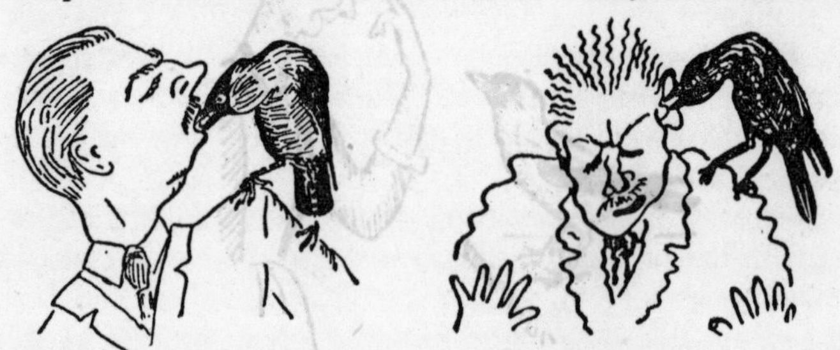

But if I did not, I had to guard my ears against him, otherwise, before I knew what was happening, the passage of one of these organs would be filled right up to the drum with warm worm pulp—for jackdaws, when feeding their female or their young, push the food mass, with the aid of their tongue, deep down into the partner's pharynx. However, this bird only made use of my ears when I refused him my mouth, on which the first attempt was always made.

* * * * *

It was entirely due to Jock that in 1927 I reared fourteen young jackdaws in Altenberg. Many of her remarkable instinctive actions and reactions towards human beings, as substitute objects for fellow-members of her species, not only seemed to fall short of their biological goal, but remained incomprehensible to me and therefore aroused my curiosity. This awakened in me the desire to raise a whole colony of free-flying tame jackdaws, and then study the social and family behaviour of these remarkable birds.

As it was out of the question that I should act as substitute for their parents and train each of these young jackdaws as I had done Jock the previous year, and as, through Jock, I was familiar with their poor sense of orientation, I had to think out some other method of confining the young birds to the place. After much careful consideration, I arrived at a solution which subsequently proved entirely satisfactory. In front of the little window of the loft where Jock had now dwelt for some time, I built a long and narrow aviary consisting of two compartments which rested upon a stone-built gutter a yard in width, and stretched almost the entire breadth of the house.

Jock was, at first, somewhat upset by the building alterations in the near neighbourhood of her home, and it was some time before she became reconciled to them and flew in and out freely through the trap-door in the roof of the front compartment of the aviary. It was only then that I proceeded to install the young birds, each of which had been made recognizable by coloured rings on one or both legs. From these rings the young jackdaws also derived their names. When the birds were all well settled in their new

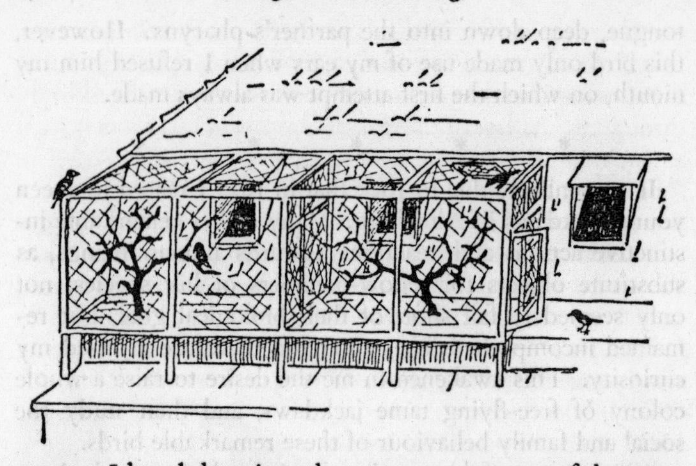

quarters I lured them into the rear compartment of the cage, leaving only Jock and the two tamest of the young birds, Blueblue and Redblue, in the front compartment, the one with the trap-door. Thus separated, the birds were again left to themselves for a few days. What I hoped to attain by these measures was that the birds destined for free flight should be held back by their social attachment to those who were still imprisoned in the hindmost part of the aviary. At this time, as I have already mentioned, Jock had begun to mother one of the young jackdaws, Leftgold, and this was very fortunate indeed as it brought about her return home at the right moment for the experiments I am about to describe. I did not choose Leftgold as one of the first subjects for release, because I hoped that, for his sake, Jock would remain in the precincts of our house, otherwise there was a risk of her flying off with Leftgold, who was now fully fledged, to live with my previously mentioned housemaid in the next village.

My hopes that the young jackdaws would fly after Jock as she had followed me were only partly fulfilled. When I opened the trap-door Jock was outside in a flash and, making one dive for liberty, within a few seconds had disappeared. It was a long time before the young jackdaws, mistrustful of the unaccustomed aspect of the open trap-door, dared to fly through it. At last both of them did so simultaneously, just as Jock came whizzing past again outside. They tried to follow her but soon lost her as neither could imitate her sharply banked turns and her steep dives. This lack of consideration for the limited flying abilities of the young is not shown by good parent jackdaws, who meticulously avoid such flying stunts while guiding their offspring. Later, when Leftgold was freed, Jock also behaved in this manner, flying slowly and refraining from all difficult manœuvring, looking back over her shoulder constantly to see whether the young bird was still following. Not only did Jock pay no attention to the other young jackdaws,

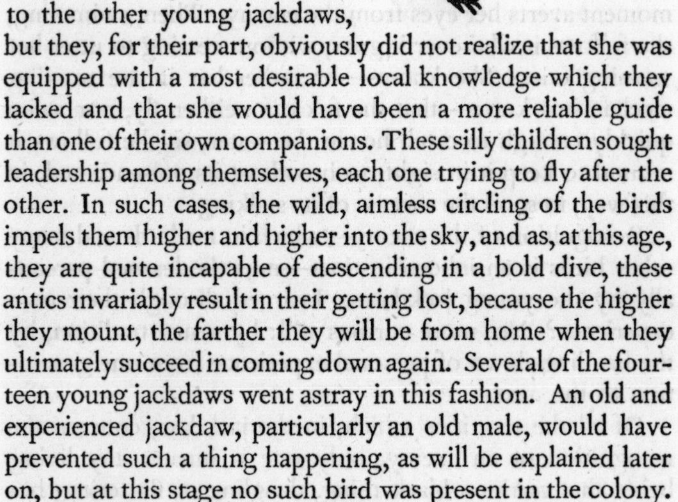

but they, for their part, obviously did not realize that she was equipped with a most desirable local knowledge which they lacked and that she would have been a more reliable guide than one of their own companions. These silly children sought leadership among themselves, each one trying to fly after the other. In such cases, the wild, aimless circling of the birds impels them higher and higher into the sky, and as, at this age, they are quite incapable of descending in a bold dive, these antics invariably result in their getting lost, because the higher they mount, the farther they will be from home when they ultimately succeed in coming down again. Several of the fourteen young jackdaws went astray in this fashion. An old and experienced jackdaw, particularly an old male, would have prevented such a thing happening, as will be explained later on, but at this stage no such bird was present in the colony.

This lack of leadership revealed itself in another and even more serious way. Young jackdaws have no innate reactions

against the enemies which threaten them, whereas a good many other birds, such as magpies, mallards or robins, prepare at once for flight at their very first sight of a cat, a fox or even a squirrel. They behave in just the same way, whether reared by man or by their own parents. Never will a young magpie allow itself to be caught by a cat, and the tamest of hand-reared mallards will instantly react to a red-brown skin, pulled along the back of the pond on a string. She will treat such a dummy exactly as if she realized all the properties of her mortal enemy, the fox. She becomes anxiously cautious and, taking to the water, never for a moment averts her eyes from the enemy. Then, swimming, she follows it wherever it goes, without ceasing to utter her warning cries. She knows—or rather her innate reacting mechanisms know—that the fox can neither fly, nor swim quickly enough to catch her in the water, so she follows it around to keep it in sight, to broadcast its presence and, in this way, to spoil the success of its stalking.

Recognition of the enemy—which in mallards and many other birds is an inborn instinct—must be learned personally by the young jackdaws. Learned through their own experience? No, more curious still: by actual tradition, by the handing-down of personal experience from one generation to the next.

Of all the reactions which, in the jackdaw, concern the recognition of an enemy, only one is innate: any living being that carries a black thing, dangling or fluttering, becomes the object of a furious onslaught. This is accompanied by a grating cry of warning whose sharp, metallic, echoing sound expresses, even to the human ear, the emotion of embittered rage. At the same time the jackdaw assumes a strange forward leaning attitude and vibrates its half-spread wings. If you possess a tame jackdaw you may, on occasion, venture to pick it up to put it into its cage or, perhaps, to cut its overgrown claws. But not if you have *two*! Jock, who was as tame as any dog, had never resented the occasional touch of my hand, but when the young jackdaws came to our house it was a different story altogether: on no account would she allow me to touch one of these small black nestlings. As, all unsuspecting, I did so for the first time, I heard behind me the sharp satanic sound of that raucous rattle, a black arrow swooped down from above, over my shoulder and on to the hand which held the jackdaw baby—astonished, I stared at a round, bleeding, deeply pecked wound in the back of my hand! That first observation of this type of attack was, in itself, illuminating as to the instinctive blindness of the impulse. Jock was, at this time, still very devoted to me and hated these fourteen young jackdaws most cordially. (Her adoption of Leftgold took place later on.) I was forced to protect them from her continually: she would have destroyed them, at one fell swoop, if she had been left alone with them for a few minutes. Nevertheless she could not tolerate my taking one of the babies into my hand. The blind reflex nature of the reaction became even clearer to me through a coincidental observation later that summer. One evening, as dusk fell, I returned from a swim in the Danube and, according to my custom, I hurried to the loft to call the jackdaws home and lock them up for the night. As I stood in the gutter, I suddenly felt something wet and cold in my trouser pocket

into which, in my hurry, I had pushed my black bathing-drawers. I pulled them out—and the next moment was surrounded by a dense cloud of raging, rattling jackdaws, which hailed agonizing pecks upon my offending hand.

It was interesting to observe the jackdaws' reaction to other black objects which I carried in my hands. My large, old, naturalist's camera never caused a similar commotion, although it was black and I held it in my hands, but the jackdaws would start their rattling cry as soon as I pulled out the black paper strips of the pack film, which fluttered to and fro in the breeze. That the birds knew me to be harmless, and even a friend, made no difference whatever: as soon as I held in my hand something black and moving I was branded as an "eater of jackdaws". More extraordinary still is the fact that the same thing may happen to a jackdaw itself: I have witnessed a typical rattling attack on a female jackdaw who was carrying to her nest the wing feather of a raven. On the other hand, tame jackdaws neither emit their rattling cry, nor make an attack, if you hold in your hand one of their own young whilst it is still naked and, therefore, not yet black. This I proved experimentally with the first pair of jackdaws which nested in my colony. The two birds, Greengold and Redgold—two of the aforementioned fourteen—were completely tame, perched on my head and

shoulders and were not in the least upset if I handled their nest and watched all their activities at close quarters. Even when I took the babies from the nest and presented them to their parents on the palm of my hand, it left them quite unmoved. But the very day that the small feathers on the nestling burst through their quills, changing their colour into black, there followed a furious attack by the parents on my outstretched hand.

After a typical rattling attack, the jackdaws are exceedingly mistrustful and hostile towards the person or animal which has given rise to it. This burning emotion stamps incredibly quickly into the bird's memory an ineradicable picture which associates the situation "jackdaw in the jaws of the enemy" with the person of the plunderer himself. Provoke a jackdaw's rattling attack two or three times running and you have lost its friendship for ever! From now on it scolds as soon as it sees you, and you are branded, even when you are not carrying a black and fluttering object in your hands. And further, this jackdaw will easily succeed in convincing all the others of your guilt. Rattling is exceedingly infectious and stimulates its hearers to attack as promptly as does the sight of the black fluttering object in the clutches of the "enemy". The "evil gossip" that you have once or twice been seen carrying such an object, spreads like wildfire, and, almost before you know it, you are notorious amongst the jackdaws in the whole district as a beast of prey which must at all costs be combated.

In most respects, all this applies equally to crows. My friend Dr Kramer had the following experience with these birds: he earned a bad reputation among the crow population in the neighbourhood of his house, by repeatedly exposing himself to view with a tame crow on his shoulder. In contrast to my jackdaws, who never resented it if one of their number perched on my person, these crows evidently regarded the tame crow sitting on my friend's shoulder as being "carried by an enemy", though it perched there of its own free will. After a short time my friend was known to all crows far and wide, and was pursued over long distances by his scolding assailants, whether or not he was accompanied by his tame bird. Even in different clothing he was recognized by the crows. These observations show vividly that corvines make a sharp distinction between hunters and "harmless" people: even without his gun, a man who has once or twice been seen with a dead crow in his hands will be recognized and not so easily forgotten.

The original value of the "rattling reaction" is doubtless to rescue a comrade from a predatory animal, or, if this is impossible, so to harry the assailant that, filled with disgust, he will renounce the hunting of jackdaws forever. Even if a goshawk or other enemy were only slightly deterred by the rattling attack from hunting jackdaws, his ensuing preference for other prey would suffice to make the jackdaws' reaction of great value for the preservation of their species. This original function of the rattling reaction is well developed in all the members of the crow family, including those species that are less gregarious than the jackdaw: similar reactions can even be found in small song-birds.

With the further development of social relations, particularly in the jackdaw, there arose in addition to the original significance of the "defence of kin reaction" a new and even more important meaning—that, through this behaviour, the recognition of a potential enemy can be communicated to the young and inexperienced birds. This is a real acquired knowledge, not a mere innate, instinctive reaction which is superficially similar to it.

I do not know whether I have made it quite clear how very remarkable all this is: an animal which does not know its enemy by innate instinct is informed by older and more experienced fellow-members of its species who or what is to be feared as hostile. This is true tradition, the handing-down of personally acquired knowledge from one generation to another. Human children might follow the example of the young jackdaws who take seriously the well-meant warnings of their parents. On the appearance of an enemy, as yet unknown to the young, an old guide jackdaw needs only to give one significant "rattle", and at once the young birds have formed a mental picture associating the warning with this particular enemy. In the natural life of jackdaws, I think it seldom happens that an inexperienced young bird first receives knowledge of the dangerous character of an enemy by seeing him with a black dangling object in his clutches. Jackdaws nearly always fly in a dense flock, in whose midst there is, in all probability, at least one bird which will begin to "rattle" at the merest sight of an enemy.

How very human this is! On the other hand, how remarkably blind and reflex-like is the innate perceptual pattern which in the inexperienced young jackdaws provokes a typical "rattling attack"! But have not we human beings also such blind, instinctive reactions? Do not whole peoples all too often react with a blind rage to a mere dummy presented to them by the artifice of the demagogue? Is not this dummy in many cases just as far from being a real enemy as were my black bathing-drawers to the jackdaws? And would there still be wars, if all this were not so?

My fourteen young jackdaws had nobody to warn them of potential dangers. Without a parent bird to give warning by rattling, such a young jackdaw will sit tight while a cat slinks up to it, or alight on the very nose of a mongrel dog, and treat him as if he were as friendly and harmless as the people in whose midst he grew up. No wonder that my jackdaw flock shrank considerably in the first weeks of its liberty. When I realized this danger and its reason I released the birds only during the hours of full daylight, at a time when few cats were abroad. The task of enticing those birds back to their cage in good time every evening occasioned me much time and trouble. "Herding a sackful of fleas"—as the German saying goes—is a trifle compared with the problem of tempting fourteen young jackdaws into an aviary. I could not touch them, for fear of starting a rattling attack, and as soon as I had manœuvred one bird, perched on my hand, through the door of the cage, two others flew out; and even if I used the foremost of the two compartments as a valve, the shutting-in process took at least an hour every evening.

The settlement of the jackdaw colony in Altenberg has cost me much work, more time—and much money, when I take into consideration the continual damage to the roof of our house. But, as I have said before, my trouble was richly rewarded. What a wonderful object for observation was this colony of completely free but absolutely trustful jackdaws! At that time—my "jackdaw time"—I knew the characteristic facial expression of every one of those birds by sight. I did not need to look first at their coloured leg-rings. This is no unusual accomplishment: every shepherd knows his sheep, and my daughter Agnes—at the age of five—knew each one of our many wild geese, by their faces. Without having known all the jackdaws personally, it would have been impossible for me to learn the inner secrets of their social life. Have you, dear reader, the slightest idea how long one must watch a flock of thirty jackdaws and how much time one must spend in close contact with them, in order to accomplish this end? It is only by living with animals that one can attain a real understanding of their ways.

Do animals thus know each other among themselves? They certainly do, though many learned animal psychologists have doubted the fact and indeed denied it categorically. Nevertheless, I can assure you, every single jackdaw of my colony knew each of the others by sight. This can be convincingly demonstrated by the existence of an order of rank, known to animal psychologists as the "pecking order". Every poultry farmer knows that, even among these more stupid inhabitants of the poultry yards, there exists a very definite order, in which each bird is afraid of those that are above her in rank. After some few disputes, which need not necessarily lead to blows, each bird knows which of the others she has to fear and which must show respect to her. Not only physical strength, but also personal courage, energy and even the self-assurance of every individual bird are decisive in the maintenance of the pecking order. This order of rank is extremely conservative. An animal proved inferior, if only morally, in a dispute, will not venture lightly to cross the path of its conqueror, provided the two animals remain in close contact with each other. This also holds good for even the highest and most intelligent of mammals. A large Nemestrinus monkey bursting with energy, owned by my friend, the late Count Thun-Hohenstein, possessed, even when adult, a deeply rooted respect for an ancient Javanese monkey of half his size, who had tyrannized him in the days of his youth. The deposing of an ageing tyrant is always a highly dramatic and usually tragic event, especially in the case of wolves and sledge-dogs, as has been observed and graphically described by Jack London in some of his Arctic novels.

The rank-order disputes in a jackdaw colony differ in one important way from those of the poultry yard, where the unfortunate Cinderellas of the lower orders eke out a truly miserable existence. In every artificial conglomeration of less socially inclined animals, such as in the poultry yard and the song-bird aviary, those higher in the social scale tend to set upon their comrades of lower rank, and the lower the standing of the individual, the more savagely will he be pecked at by all and sundry. This is often carried so far that the wretched victim, bullied from all sides, is never able to rest, is always short of food and, if the owner does not interfere, may finally waste away altogether. With jackdaws, quite the contrary is the case: in the jackdaw colony those of the higher orders, particularly the despot himself, are not aggressive towards the birds that stand far beneath them: it is only in their relations towards their immediate inferiors that they are constantly irritable; this applies especially to the despot and the pretender to the throne—Number One and Number Two. Such behaviour may be difficult for a casual observer to understand. A jackdaw sits feeding at the communal dish, a second bird approaches ponderously, in an attitude of self-display, with head proudly erected, whereupon the first visitor moves slightly to one side, but otherwise does not allow himself to be disturbed. Now comes a third bird, in a much more modest attitude which, surprisingly enough, puts the first bird to flight; the second, on the other hand, assumes a threatening pose, with his back feathers ruffled, attacks the latest comer and drives him from the spot. The explanation: the latest comer stood in order of rank midway between the two others, high enough above the first to frighten him and just so far beneath the second as to be capable of arousing his anger. Very high-caste jackdaws are most condescending to those of lowest degree and consider them merely as the dust beneath their feet; the self-display actions of the former are here a pure formality and only in the event of too close approximation does the dominant bird adopt a threatening attitude, but he very rarely attacks.

The degree of animosity of the higher orders towards the lower is in direct proportion to their rank, and it is interesting to note that this essentially simple behaviour results in an impartial levelling-up of the disputes between individual members of the colony. The gestures of anger and attack may also stimulate those against whom they are not directed. I myself, when I hear two people cursing each other in an overcrowded tramcar, have to suppress an almost uncontrollable desire to box the ears of both parties soundly. High-ranking jackdaws evidently feel the same emotion, but, as they are in no way inhibited by the horror of making a scene, they interfere vigorously in the quarrel of two subordinates as soon as the argument gets heated. The arbitrator is always more aggressive towards the higher-ranking of the two original combatants. Thus a high-caste jackdaw, particularly the despot himself, acts regularly on chivalrous principles—where there's an unequal fight, always take the weaker side. Since the major quarrels are mostly concerned with nesting sites (in nearly all other cases the weaker bird withdraws without a struggle), this propensity of the strong male jackdaws ensures an active protection of the nests of the lower members of the colony.

Once the social order of rank amongst the members of a colony is established it is most conscientiously preserved by jackdaws, much more so than by hens, dogs or monkeys. A spontaneous reshuffling, without outside influence, and due only to the discontent of one of the lower orders, has never come to my notice. Only once, in my colony, did I witness the dethroning of the hitherto ruling tyrant, Gold-green. It was a returned wanderer who, having lost in his long absence his former deeply imbued respect for his ruler, succeeded in defeating him in their very first encounter. In the autumn of 1931 the conqueror, "Double-aluminium"—he derived this strange name from the rings on his feet—came back, after having been away the whole summer. He returned home strong in heart and stimulated by his travels, and at once subdued the former autocrat. His victory was remarkable for two reasons: first, Double-aluminium, who was unmated and therefore fighting alone, was opposed in the struggle by both the former ruler and his wife. Secondly, the victor was only one and a half years old, whereas Goldgreen and his wife both dated back to the original fourteen jackdaws with which I started the settlement in 1927.

The way in which my attention was drawn to this revolution was quite unusual. Suddenly, at the feeding-tray, I saw, to my astonishment, how a little, very fragile, and, in order of rank, low-standing lady sidled ever closer to the quietly feeding Goldgreen, and finally, as though inspired by some unseen power, assumed an attitude of self-display, whereupon the large male quietly and without opposition vacated his place. Then I noticed the newly returned hero, Double-aluminium, and saw that he had usurped the position of Goldgreen, and I thought at first that the deposed despot, under the influence of his recent defeat, was so subdued that he had allowed himself to be intimidated by the other members of the colony, including the aforesaid young female. But the assumption was false: Goldgreen had been conquered by Double-aluminium only, and remained forever the second in command. But Double-aluminium, on his return, had fallen in love with the young female and within the course of two days was publicly engaged to her! Since the partners in a jackdaw marriage support each other loyally and bravely in every conflict, and as no pecking order exists between them, they automatically rank as of equal status in their disputes with all other members of the colony; a wife is therefore, of necessity, raised to her husband's position. But the contrary does not hold good—an inviolable law dictates that no male may marry a female that ranks above him. The extraordinary part of the business is not the promotion as such but the amazing speed with which the news spreads that such a little jackdaw lady, who hitherto had been maltreated by eighty per cent of the colony, is, from to-day, the "wife of the president" and may no longer receive so much as a black look from any other jackdaw. But more curious still—the promoted bird knows of its promotion! It is no credit to an animal to be shy and anxious after a bad experience, but to understand that a hitherto existent danger is now removed and to face the fact with an adequate supply of courage requires more sense. On a pond, a despot swan rules with so tyrannical a rule that no other swan, except the wife of the feared one, dares to enter the water at all. You can catch this terrible tyrant and carry him away before the eyes of all the others, and expect that the remaining birds will breathe an audible sigh of relief and at once proceed to take the bath of which they have been so long deprived. Nothing of the kind occurs. Days pass before the first of these suppressed subjects can pluck up enough courage to indulge in a modest swim hard against the shores of the pond. For a much longer time, nobody ventures into the middle of the water.

But that little jackdaw knew within forty-eight hours exactly what she could allow herself, and I am sorry to say that she made the fullest use of it. She lacked entirely that noble or even blasé tolerance which jackdaws of high rank should exhibit towards their inferiors. She used every opportunity to snub former superiors, and she did not stop at gestures of self-importance, as high-rankers of long standing nearly always do. No—she always had an active and malicious plan of attack ready at hand. In short, she conducted herself with the utmost vulgarity.

You think I humanize the animal? Perhaps you do not know that what we are wont to call "human weakness" is, in reality, nearly always a pre-human factor and one which we have in common with the higher animals? Believe me, I am not mistakenly assigning human properties to animals: on the contrary, I am showing you what an enormous animal inheritance remains in man, to this day.

And if I have just spoken of a young male jackdaw falling

in love with a jackdaw female, this does not invest the animal with human properties, but, on the contrary, shows up the still remaining animal instincts in man. And if you argue the point with me, and deny that the power of love is an age-old instinctive force, then I can only surmise that you yourself are incapable of falling a prey to that passion.

A strange thing, this "falling in love". The metaphor expresses the psychical process with a drastic sense of realism—an audible bump, and you are in love! It would be impossible to symbolize it more aptly. And in this connection, many higher birds and mammals behave in exactly the same way as the human being. Very often even in jackdaws the "Grand Amour" is quite suddenly there, from one day to the next—indeed most typically, just as in the case of man, at the moment of the first encounter. Marlowe says:

> The reason no man knows; let it suffice,
> What we behold is censured by our eyes.
> Where both deliberate, the love is slight;
> Who ever loved that loved not at first sight?

This famous "love at first sight" plays a big role in the life of wild geese and jackdaws, and this may be most impressive for the observer. I know of a number of cases where love and troth were plighted on the occasion of the first meeting. The continual presence of the loved one is not so conducive to this state of mind as one might at first imagine. It can even be disadvantageous. At any rate, a temporary parting may achieve that which was hindered by years of intimacy. In the case of wild geese, I have repeatedly noticed that a betrothal was pledged when two fairly close friends met again after a fairly long separation. Even I myself have been affected by this quite typical phenomenon—but that is another story.

Many of my readers, particularly those with some psychological education, will have raised their eyebrows critically at the word "betrothal": it is customary to consider the animal as more or less "bestial", and to believe that love and marriage in animals are governed by much more sensual impulses than in man. This idea is quite wrong in the case of those animals in whose life love and marriage play a major part. Amongst those few birds which maintain a lasting conjugal state, and whose behaviour in this respect has been explored to the very last detail, the betrothal nearly always precedes the physical union by quite a long period of time. In those species which marry only for one brood, as for example most small song-birds, herons and many others, the engagement is necessarily of shorter duration. But nearly all those that marry for life become "engaged" long before they "wed". The record for long engagements, in small birds, is held by the bearded tit, to which my friends Otto and Lilli Koenig dedicated years of observation and one of their most delightful books. These beings—I mean the tits and not the Koenigs—become engaged, strangely enough, in their juvenile plumage, before their first moult, at the age of two and a half months, that is to say about nine months before they are sexually mature and mate for the first time. To the connoisseur this is something quite remarkable. The unique display ceremonies, especially the courtship-actions of the male, are calculated to expose the wonderful details of his mature plumage, above all his black "mutton-chop" whiskers and the deep ebony of his lower tail coverts. The little fellow shows off a beard and unfolds tail feathers whose conspicuous colouring will not become evident until two months later. Of course he does not "know" what he looks like, and the innate movements are intended for the finished adult plumage only. The autumn betrothals of surface-feeding ducks are a different matter. The drakes are at this time as incapable of reproduction as the young bearded tits, but strut already in their full gala dress which does not change till after mating time in early spring.

Jackdaws, like wild geese, become betrothed in the spring following their birth, but neither species becomes sexually mature till twelve months later; thus the normal period of betrothal is exactly a year. The wooing of the male jackdaw is so far similar to that of the gander and the young human male in that none of these has at its disposal particular instruments of courtship: they cannot spread the splendour of a peacock's tail, nor, like Shelley's skylark, pour forth their "full heart in profuse strains of unpremeditated art". The "eligible" jackdaw must make the most of himself without any of these accessories, and the way he does so is astonishingly human. Exactly as the greylag gander, so the young jackdaw "spreads himself" to denote his superfluous energy. All his movements are consciously strained, and his proudly reared head and neck are held in a permanent state of self-display. He provokes the other jackdaws continually if "she" is looking, and he purposely becomes embroiled in conflicts with otherwise deeply respected superiors.

Above all, he seeks to impress his loved one with the possession of a potential nesting cavity, from which he drives all other jackdaws, irrespective of their rank, at the same time giving utterance to the high, sharp "zick, zick, zick" of his nesting call. This calling-to-nest ceremony is, for the moment, purely symbolic. At this stage it is immaterial whether the cavity in question is really suitable for a nest. In contrast to that of the jackdaw the parallel ceremony in the house sparrow is to be taken seriously: the male house sparrow only thinks of marrying when he has found and fought for what he considers an adequate nest cavity, for which there is always a wild "scrum" amongst male sparrows. For the "Zick ceremony" of the jackdaws, any dark corner or small hole, too narrow to be crawled into, serves the purpose. The already-mentioned male jackdaw who used to stuff my ears with mealworms showed a preference for zicking on the edge of a very small mealworm pot. Our free-living jackdaws use, for the same purpose, the upper opening of our chimney-pots, although they rarely nest there, and their muffled "zick, zick" can be heard in spring-time from the various stoves in our living-rooms.

All these different forms of self-presentation are addressed by the courting male always to one special female. But how does she know that the whole act is being performed for her benefit? This is all explained by the "language of the eyes", which Byron, in *Don Juan*, calls:

> The answer eloquent where the soul shines,
> And darts in one quick glance a long reply.

As he makes his proposals the male glances continually towards his love, but ceases his efforts immediately if she chances to fly away; this, however, she is not likely to do if she is interested in her admirer.

Remarkable and exceedingly comical is the difference in eloquence between the eye-play of the wooing male and that of the courted female: the male jackdaw casts glowing glances straight into his loved one's eyes, while she apparently turns her eyes in all directions other than that of her ardent suitor. In reality, of course, she is watching him all the time, and her quick glances of a fraction of a second are quite long enough to make her realize that all his antics are calculated to inspire her admiration; long enough to let "him" know that "she" knows. If she is genuinely not interested, and will not look at him at all, then the young

jackdaw male gives up his vain efforts as quickly as—well, any other young fellow. To her swain, now proudly advancing in all his glory, the young jackdaw lady finally gives her assent by squatting before him and quivering, in a typical way, with her wings and tail. These movements of both partners symbolize a ritual mating invitation, though they do not lead to actual union but are purely a greeting ceremony. Married jackdaw ladies greet their husbands in the same way, even outside the mating season. The purely sexual meaning attached to this ceremony in the genealogy of the species has been entirely lost and it now only serves to signify the affectionate submission of a wife to her husband. It corresponds in its meaning almost exactly with "symbolic inferiorism" in fishes. From the moment that the bride-to-be has submitted to her male, she becomes self-possessed and aggressive towards all the other members of the colony. For a female, the betrothal entails a high promotion in the colony, for being, on the average, smaller and weaker than the male, she stands much lower in rank than he as long as she is single.

The betrothed pair form a heart-felt mutual defence league, each of the partners supporting the other most loyally. This is essential, because they have to contend with the competition of older and higher-standing couples in the struggle to take and hold a nesting cavity. This militant love is fascinating to behold. Constantly in an attitude of maximum self-display, and hardly ever separated by more than a yard, the two make their way through life. They seem tremendously proud of each other, as they pace ponderously side by side, with their head feathers ruffled to emphasize their black velvet caps and light grey silken necks. And it is really touching to see how affectionate these two wild

creatures are with each other. Every delicacy that the male finds is given to his bride, and she accepts it with the plaintive begging gestures and notes otherwise typical of baby birds. In fact, the love-whispers of the couple consist chiefly of infantile sounds, reserved by adult jackdaws for these occasions. Again, how strangely human! With us too, all forms of demonstrative affection have an undeniable child-like tendency—or have you never noticed that all the nick-names we invent, as terms of endearment for each other, are nearly always diminutives?

The male jackdaw's habit of feeding his wife is a charming gesture which appeals directly to our human understanding, and the chief expression of tenderness shown by the female is no less attractive to our minds. It consists in her cleaning those parts of his head feathers which he cannot reach with his own bill. Friendly jackdaws, as also many other social birds and animals, often perform mutually the duty of "social grooming", without any ulterior erotic motive. But I know of no other being which so throws its heart into the process as a lovesick jackdaw lady! For minutes on end —and that is a long time for such a quicksilvery creature—she preens her husband's beautiful, long, silken neck feathers, and he, with sensuous expression and half-shut eyes, stretches his neck towards her. Not even in the proverbial doves or

love-birds does the tenderness of married love find such charming expression as in these notorious corvines! And the most appealing part of their relationship is that their affection increases with the years instead of diminishing. Jackdaws are long-lived birds and become nearly as old as human beings. (Even small birds like warblers or canaries live almost two decades and are still capable of reproduction at the age of fifteen or sixteen years.) Now jackdaws, as described, become betrothed in their first year, and marry in their second, so their union lasts long, perhaps longer than that of human beings. But even after many years, the male still feeds his wife with the same solicitous care, and finds for her the same low tones of love, tremulous with inward emotion, that he whispered in his first spring of betrothal and of life. You may not believe it, but there are other animals in whom—though they may live in life-long marital union—the situation is different: in whom the glowing fires of the first season of love become extinguished by cool habit; with whom the thrilling enchantment of courtship's phrases entirely disappears as time goes on: and in whose further mutual association all the activities of wedlock and family life are performed with the mechanical apathy common to other everyday practices.

Of the many jackdaw betrothals and marriages whose

course I was able to follow, only one disintegrated, and that was during the period of betrothal. The cause of the trouble was a young jackdaw lady of unusually vivacious temperament, called Leftgreen, whose romance, with its happy end, was the diametrical opposite of the tragic love affair of the greylag goose Maidy, of whom I shall tell you in another book. In the early spring of 1928, which was the first spring in the life of my first "Fourteen", the reigning despot, Goldgreen, pledged his troth with Redgold, who was obviously the fairest of the eligible virgins: indeed, had I been a jackdaw I would have chosen her myself. The second jackdaw of the colony, Bluegold, had also made patent overtures to Redgold, but he soon relinquished them and became engaged to Rightred, a rather big and, for a female, strongly built jackdaw. This betrothal ran a slower and less thrilling course than that of Goldgreen and Redgold, and was obviously a more lukewarm affair than the "grand passion" of the latter couple.

Leftgreen was at this time—at the beginning of April—not even "boy-conscious", for the awakening of sexual activity in year-old jackdaws varies considerably in the time of its commencement. It was not till the beginning of May that Leftgreen appeared on the scene, and her entry was as impulsive as it was sudden. As I have said before, she was small, and low in the order of rank; and, from a human point of view, she was much less lovely than Rightred, to say nothing of Redgold. But there was something about her . . . She fell in love with Bluegold, and her love was so much more ardent than that of Rightred that—to begin with the end, in anything but logical style—she finally outwitted her stronger and more beautiful rival.

The first sign which I received of the impending love drama was the enacting of the following scene. Bluegold sat peacefully on the upper edge of the open aviary door and allowed Rightred, who was sitting on his left side, to preen

his neck feathers. Suddenly, unnoticed by both, Leftgreen also landed on the door and sat for a time about a yard away, casting on the lovers glances rife with tension. Then, slowly and carefully, she sidled, from the right, ever closer to Bluegold, and with outstretched neck and, as a measure of caution, wings prepared for flight, she, too, began to caress his neck feathers. Bluegold did not notice that his toilet was being effected from both sides, having closed both his eyes in complete abandonment to the pleasures of the process. Rightred was also quite oblivious of her rival's presence, since between her and Leftgreen was interposed the large fat form of her fiancé, now made even bigger by his fully ruffled feathers. This tense situation prevailed for some minutes, until finally Bluegold happened to open his right eye, and when he saw the strange female at such close quarters he pecked at her with sudden vehemence. At the same moment Leftgreen was discovered by Rightred also, whose line of vision became suddenly cleared by the altered position of the angry male. With one bound, she leapt over

her betrothed and threw herself with such fury upon her rival that I received the impression that, unlike me, Rightred was already well aware of the earnest intentions of little Leftgreen. The rightful bride seemed fully alive to the seriousness of the situation; never, before or since, have I seen one jackdaw pursue another with such venom. But she had no success. The smaller and sprightlier Leftgreen surpassed her in the art of flying, and when Rightred, after a long air hunt in pursuit of her detested rival, landed at the side of her betrothed, she was completely out of breath; the little Leftgreen, on the other hand, who arrived not a minute later, seemed quite collected. And that settled the matter! In her importunate courtship, Leftgreen was admirably tenacious rather than subtle: she pursued the couple day after day without the slightest pause, whether they walked or flew, but kept just far enough away to avoid unduly provoking them. But as soon as the pair nestled close together in homely comfort, Leftgreen approached nearer and watched patiently for the moment when Rightred should softly scratch the head of her lover.

Many drops wear out the stone. The attacks of Rightred lost gradually in intensity, Bluegold ceased to object to the bilateral attentions, and one day I noticed that things had reached the following pass. Bluegold was sitting still, letting Rightred tickle the back of his head. On the other side, Leftgreen proceeded to do the same thing. Then, for some reason, Rightred stopped scratching and flew away. The big male opened his eyes and beheld Leftgreen on his other side. But did he peck her? Did he drive her away? No! Pensively turning his head, he deliberately offered to Leftgreen the coveted part of the nape of his neck! Then his eyes closed again.

From now on, Leftgreen gained rapidly in his favour. A few days later I saw that he was feeding her, regularly and tenderly, though, to be sure, in the absence of Rightred:

not that he was consciously doing this behind the back of his "rightful" bride—to believe this would be to overrate the mental capacity of the jackdaw. Had Rightred been present, she would undoubtedly have received the delicacy, but because she was not, the other obtained it. My friend A. F. J. Portieje observed similar behaviour in mute swans. An old, married swan male furiously expelled a strange female, who came close to the nest where his wife was sitting and made him proposals of love. But on the very same day he was seen to meet this new female, remote from his wife and his nest, on the other side of the lake, and to succumb to her temptations without further ado. Here, too, a human parallel can be found, but here again it is erroneous. In the precincts of his nest, the male swan is concerned primarily with the defence of his territory, and he sees in every strange member of his species, whether it be male or female, only the intruder. Away from his nest territory where all trespassers must be prosecuted, he is not thus preoccupied, and is, therefore, able to recognize the desirable female in the person of the newcomer.

The surer Leftgreen became of the male, the more impudent she became towards Rightred. No longer did she flee her rival, and there were sometimes duels between the two females. Strange was the behaviour of Bluegold in this dilemma. Whereas normally he had bravely supported his wife in any quarrels with other members of the colony, now he was obviously in conflict with himself. He certainly threatened Leftgreen, but never more took action against her. Indeed, I once saw him make slight threatening gestures towards Rightred, and his inhibitions and embarrassment in this situation were often quite apparent.

The end of this romance was sudden and dramatic. One fine day Bluegold had disappeared, and with him—Leftgreen! I could not assume that these two mature and experienced birds had met with an accident at the same moment;

doubtless they had flown away together. Conflicting emotions are certainly just as tormenting to animals as to human beings, and of this I will speak later; and I cannot exclude the possibility that it was these irreconcilable sentiments that impelled the male jackdaw to leave the colony.

I have never known an occurrence of this type in older nesting couples, and I do not think that such a thing ever happens. All the jackdaw nesting pairs that I was able to observe for any length of time remained true to each other to the day of their death. Nevertheless, widows and widowers remarry without demur, as soon as they find a suitable partner, though this is not so easy for old and high-caste females. Greylag geese never remarry, and this is a subject which I have treated in my book about these birds.

In their second year jackdaws become capable of reproduction. In reality, they probably are so in their second autumn, immediately after their first full moult in which not only their body plumage but also the large flight feathers of their wings and tail are renewed. After this moult, on fine autumn days, the birds are obviously in a mood for sexual activity, and are especially inclined to seek nesting cavities. The aforementioned "zick, zick" can be heard continuously from all sides. When the weather becomes cooler this post-moulting sexual mood fades out again, but remains latent, and on warm winter days little zick-zick concerts sometimes ring through the chimneys into the rooms below. In February and March the matter becomes serious and the "zick, zick" hardly ever ceases during the hours of daylight.

At this time another ceremony is often performed which is quite the most interesting in the whole social life of the jackdaws. In the last days of March when the zicking has reached its height, the concert swells, in some niche in the wall, to an unprecedented volume. At the same time it alters its timbre, becomes deeper and fuller and sounds, from now on, more like a "yip, yip, yip" emitted in an ever-increasing succession of rapid staccato notes which, towards the end of the strophe, reaches a pitch of frenzy. Simultaneously, excited jackdaws rush in from all sides towards this niche and, with ruffled feathers and their best threatening attitudes, join in the yipping concert.

And what does all this mean? It took me quite a time to find out: it represents neither more nor less than communal action against a social delinquent! In order fully to understand this collective reaction, which is purely instinctive, we must look further into it.

In general, a jackdaw zicking in its nesting cavity will not lightly be attacked, as the aggressor will inevitably be at a disadvantage. Now the jackdaw has *two* separate ways of threatening, as distinct in their form as in their meaning: should the quarrel concern exclusively a social rank dispute, the rivals threaten each other by drawing themselves up to their full height and flattening their feathers. This attitude implies the intention of flying upwards and onto the back of the adversary. It is the forerunner of that method of fighting, common also to cocks and many other birds, in which the partners fly upwards, locked in fight, each clawing and striking at his opponent, endeavouring to overcome him and to throw him over on his back. The second threatening attitude is exactly the opposite. The bird ducks, drawing

low his head and neck, to form a curious "cat's back", emphasized by the ruffling of his back feathers. The tail is drawn sideways toward the opponent and spread into a fan. While in the first threatening attitude the bird tries to appear as *tall* as possible, in this, the second one, he makes himself as *bulky* as he can. The first attitude means, "If you don't make room, I shall attack you flying", while the second implies, "Here where I sit I will fight to the last, for I shall not cede one inch"! A bird of high rank which approaches a subordinate, with the intention of driving him from a particular place, generally retires if the latter assumes the second threatening attitude. Only if the aggressor himself sets store by this spot, for example with a view to a nesting-site, does he proceed to further action. In this case he also assumes the second attitude. And so the two squat for a long while, shoulder to shoulder, each watching the other with grim intensity. They hardly ever come to blows but, still squatting in the same place and keeping their distance, aim fast and furious but totally ineffective pecks at each other. The sharp expulsion of breath and the snapping of the beak is distinctly audible at each peck. The result of such quarrels is always a question of who can hold out the longest.

The whole zicking ceremony is bound up inseparably with the second attitude of threatening, the jackdaw being quite unable to utter either its "zick, zick" or its "yip, yip" in any other position. In the jackdaw, as in all animals which mark our territories, the boundary between the "possessions" of two rivals is determined by the fact that any individual will fight much more furiously when near its home, than when it is on foreign ground. Therefore a jackdaw zicking in its own rightful nesting cavity has from the start a very appreciable advantage over every intruder, and this superiority, as a rule, more than outweighs any difference in strength and rank that might exist between fellow-members of the colony.

However, owing to the keen competition for the possession of serviceable nest cavities, it sometimes happens that a very strong bird attacks a very much weaker one in its nesting cavity and assaults it unmercifully. In this eventuality, what I have called the "yip-reaction" comes into play. The zicking of the outraged householder increases at first tremendously and then gradually gives place to a yipping. If his wife is not already at hand to assist, she now comes rushing up with ruffled feathers, joins in the yipping and attacks the peace-breaker. Should the latter not retreat instantly the incredible happens: loudly yipping, all the jackdaws within earshot storm the threatened nest cavity and the original battle disappears in a solid mass of jackdaws, in an increasing paroxysm of rage, a crescendo, accelerando and fortissimo of general yipping. After thus forcefully discharging their excitement, the birds disperse soberly; only the nest-owners can still be heard quietly zicking in their once more peaceful home.

The congregation of a number of jackdaws is usually enough to terminate the fight, particularly since the original aggressor participates in the yipping! This might seem to an observer, who attributed to the bird human qualities, that the cunning invader wished to divert suspicion from himself by crying "Stop thief". In reality, however, the aggressor, dragged willy-nilly into the yipping reaction, does not even know that he is the cause of the tumult. And so, yipping, he turns in all directions as though he, too, were seeking the culprit, and, strangely enough, he is doing so in all sincerity.

I have often seen cases, however, where the aggressor was very definitely recognized by the advancing members of the colony, and was thoroughly thrashed by them if he persisted in his attack. In 1928 the real despot of the jackdaw colony was a saucy magpie whom I had reared together with the jackdaws. The magpie far surpasses the jackdaw in fighting qualities and, unlike it, is a distinctly non-social bird and quite devoid of the finely adjusted social drives and inhibitions which make the jackdaw so appealing to us. So this feathered rascal, lacking any sense of propriety, soon became the same disturbing element in the jackdaw colony as the inveterate criminal in a civilized human society. Time and again this piebald bully forced his way into the nesting cavities of different jackdaw couples and incited an indignant yipping. Although the magpie had no organ for the yipping reaction of the jackdaws and pursued his object undaunted, he was nevertheless forcefully brought to his senses by the mass attack of the jackdaws, who taught him, by bitter experience, to leave their nests alone. Thus, contrary to my earnest fears, eggs and young came off unscathed.

It is primarily the old, strong, high-ranking males that play the most important role in the yipping and rattling reactions. In another way also they safeguard the welfare of the community. In the autumn of 1929 a huge flock of migrating jackdaws and rooks, all in all about two hundred, descended on the fields in the immediate neighbourhood of our house. And all my young jackdaws of that year and the previous one got themselves inextricably mixed up with these strangers. Only my few old birds stayed at home. I regarded this occurrence as an absolute catastrophe and visualized my work of two years flying away beyond recall. I knew only too well how strong an attraction a migrating flock can exert upon young jackdaws, who, intoxicated by the sight of a myriad ebony wings, seem compelled to fly with them; and, had it not been for Goldgreen and Bluegold, the results of my hard labour would have gone with the wind (or rather, against it, since jackdaws prefer to fly in that direction). These two old males, the only ones of their age in the colony, flew incessantly backwards and forwards between house and field, and there they did something so incredible that I should be inclined to doubt it myself as I write, had the same type of activity not frequently been witnessed and even experimentally proved by myself and my workers. Each of the two patriarchs sought, from out of the crowd, a single one of our own young birds and fetched it home in a most peculiar manner. They induced it to take wing by a very special manœuvre which jackdaw parents also practise with their children when enticing them from some dangerous place. The parent bird flies, from behind, low over the back of the young one and, the moment he is immediately above him, he executes a quick sideways

wobbling movement with his closely folded tail, which ceremony impels the sitting bird to "follow the leader" with a reflex-like certainty. This feat being achieved, the old males resumed their homeward flight, ever casting backward glances to see if their charges were still following. We have already seen how Jock also used this method in guiding her wards.

During the whole procedure, Goldgreen and Bluegold uttered continuously a significant call-note, clearly distinguishable from the usual short, light flight-call by the drawn-out length of its dark, muffled tone. While the ordinary flying-call sounds like a high "Kia, Kia", this second note can be expressed by a "Kiaw, Kiaw". I was conscious at once that I had already heard this cry, but only now was its meaning brought home to me.

The two male jackdaws worked with a feverish haste; well-trained sheepdogs, who separate and round up their own sheep from a large flock, could not have shown a keener efficiency. They worked without pause, well into the dusk, when jackdaws have normally long since sought their perches. Theirs was no easy task, for no sooner had they coaxed a contingent of young jackdaws into their home, than these immediately flew off to rejoin the flock on the meadow: for every ten birds that were laboriously recaptured, nine escaped again. But late in the evening, when the wandering tribe moved onward, I found with a deep sigh of relief that, of our many young jackdaws, only two were missing.

Impressed by this episode, I began to investigate more thoroughly the different meanings of "Kia" and "Kiaw". They were soon clear to me: both calls denote "Fly with me!" But whereas the jackdaw calls "Kia" when it is in the mood for flying abroad, it cries "Kiaw" to express a homeward-bound intention. I had always noticed that migrating flocks of jackdaws cried differently, more shrilly than my birds, and I now know the reason why. Far from home, with all the ties of the "homing instinct" temporarily severed, the motivation for the "Kiaw" call is absent. Under these conditions, only the wander-call "Kia" is heard. In this connexion, it would be interesting to ascertain whether the "Kiaw" sound is ever heard in the spring in flocks migrating back to their breeding colonies. What I heard from my own jackdaw flock was invariably a mixture of both notes because, within the precincts of the colony, a certain homeward-bound tendency was never entirely lacking, even in winter.

Despite the verbal interpretation "Fly with me", it must be stressed that these call-notes are purely indicative of the mood of the bird in question and are in no way a conscious command. But these completely unintentional expressions of individual feeling are of as highly infectious a nature as yawning in human beings. It is this mutual mood-infection which ensures that all the jackdaws finally act concertedly. Thus, far from being determined by the authority of an autocratic leader, the activities of flocks of birds, herds or packs of animals, and even schools of fishes, are decided by a process very similar to the democratic system of voting.

This is the reason why the behaviour of a flock of jackdaws under certain circumstances shows a regrettable lack of unison. This interaction of moods may sometimes continue for a surprisingly long period, thus emphasizing the birds' utter inability to come to a decision: this would involve the aptitude to concentrate on one particular motive by consciously subjugating all other present impulses: but this faculty is an attribute only of man and, to a much lesser extent, some of the more highly intelligent mammals. It makes a human observer positively nervy to watch such a jackdaw flock torn hither and thither between "Kia" and "Kiaw" calls, for half-hours on end. There sits the flock, in the middle of a field, some miles away from home; it has relinquished its quest for food, so the birds will soon be flying home, "soon" meaning of course a jackdaw's somewhat elastic idea of that conception of time. At last a few birds—usually old, strongly reactive ones—take off, emitting "Kiaw" cries and thereby provoking the whole flock to leave the ground with them; but no sooner are they in the air than it becomes evident that many members of the flock are still in "Kia" mood. In a babel of "Kia" and "Kiaw" cries, the flock circles and eventually lands again, this time on a field perhaps farther still from home. This is repeated a dozen times, then very gradually the "Kiaw" factor becomes preponderant, gains ascendancy and finally sweeps all before it with the voracity of an avalanche.

The "Kiaw" reaction certainly plays an overwhelmingly large role in maintaining the integrity of the colony. I have already related how, on one occasion, it preserved mine from ruin and, later on, it did so again in an entirely different manner. Some years after its establishment my jackdaw colony was struck by a catastrophe whose cause remains obscure to this day. In order to avoid the inevitable losses of winter migration, I kept the birds confined to the aviary from November to February and paid an assistant, who was said to be conscientious, to look after them, as I was living at this time in Vienna. One day all the birds had gone! The wire of the cage had a hole in it, possibly caused by the wind, two jackdaws were found dead and the rest had disappeared. Perhaps a marten had got in, but I never found out. Keepers of free-living animals become accustomed to all sorts of set-backs, but this loss was perhaps "the most unkindest cut of all" that my tireless efforts in animal-rearing ever received. But it brought some good withal, in the form of some observations which would otherwise never have been possible. This luck began with the sudden reappearance of one bird after the space of three days: it was Redgold, the ex-queen, the first jackdaw who had hatched and brought up her young in Altenberg.

This lone jackdaw rarely ventured forth but sat the live-long day on the weathercock—and sang! She sang almost without intermission! All song-birds, to which group the corvines also belong, tend to sing profusely when in solitude or robbed of the opportunity to pursue their normal activities, in other words when they are "bored". For this reason, the bird kept in solitary confinement sings much more than the one which enjoys its freedom. All the energies which would otherwise be disseminated over a hundred and one different activities flow concurrently in the one channel of the song. In nature, also, where the song of most small singing birds serves to mark the boundaries of territorial rights and intimates an invitation to the female, those males who have found no mate sing louder and longer than their happily mated brothers. Because of the predominance of males, many must remain celibate, but this does not appear to depress them unduly. Contrary to the opinion of our societies for prevention of cruelty to animals, it is therefore no great act of cruelty to keep a nightingale or a goldfinch alone in captivity for the purpose of its song, and Blake's adage:

> A Robin-redbreast in a cage
> Puts all Heaven in a rage

need not be taken too seriously. The male lapdog on one end of the lead and the frustrated spinster on the other are objects far more deserving of our pity. Speaking for myself, however, I must admit that the continuous song of singly kept birds gets on my nerves, with time. My male common redstart, who rarely sings, as he lives in a large cage with his wife, but who, as I write these lines, is performing his remarkable courtship dance in front of the lady of his heart, affords me much more pleasure than the most voluble of solitary singers. All the same, the singly kept male song-bird does not suffer, nor is his song the expression of sorrow and desire, as sentimental people like to believe. If he is at all distressed his carolling ceases at once.

But the lonely jackdaw lady, Redgold, was genuinely sad, and I am not anthropomorphizing when I say that she suffered mentally. Mental suffering in animals is practically always dumb, but in this one case—I know of no other—her sorrow found vocal outlet, intelligible to man, or at least to one man who understood "Jackdawese". The song of all jackdaws—for both sexes sing equally well—consists of an infinite variety of notes, both specific and mimicked, and this motley of sound is woven to a design which, though not beautiful, is a comfortable and homely singsong. In the jackdaw, the mimicking or so-called mocking of other sounds does not play a marked role and is not nearly so perfected as in the crow and the raven; nevertheless, singly kept jackdaws learn to imitate human words quite well. But a very curious speciality of their song is a phenomenon which one might interpret as self-mimicking. In the songs of the jackdaw each and all of the different cries peculiar to the species are constantly reiterated. All the call-notes with which we have already become acquainted are reproduced in the song, and that includes the "Kia" and "Kiaw" cries, "zicking" and "yipping" and even the sharp rattle normally used in defence of a comrade. In all other birds that I know, sounds with a "meaning" are not used in the song at all, or, at the most, they occur only singly. But the song of free-living jackdaws consists almost entirely of such sounds! And the unique part of it is that the singer accompanies the individual cries with the corresponding gestures. When rattling he bends forward and quivers with his wings, just as in a genuine rattling reaction; when "zicking" or "yipping" he assumes the appropriate threatening attitude. In other words he behaves exactly as a human being who becomes so engrossed in the recitation of a ballad that the individual passages awaken corresponding feelings and emotions and automatically evoke the appropriate gestures. To my human ear, these "song-sounds" are in no way distinguishable from those which are meant in earnest. How often have I rushed, in alarm, to the window, hearing a loud rattling and thinking a marauder had one of my birds in its clutches, only to find that a loudly reciting jackdaw had made a fool of me. But never have I seen a real jackdaw taken in, in that way. This is a constant source of wonderment to me, considering the blind, reflex-like nature of the reaction which follows on the rattling of a fellow-member of the species in cases of emergency. It is this significance of the individual sounds and still more the touching expressiveness of the accompanying gestures that make the jackdaw's song so enchanting to one who understands its emotional movements and sounds. How delightful are these little black fellows, repeating with elation their ballads, in which are conjured up pictures of all the exciting experiences pertaining to the life of a jackdaw!

But the song of the lone jackdaw Redgold was really heart-rending. It was not how she sang, but what she sang. Her whole song was suffused with the emotion which obsessed her, with the sole desire of bringing back her lost ones by means of the "Kiaw" call, "Kiaw" and again "Kiaw" in all tones and cadences, from the gentlest piano to the most desperate fortissimo. Other sounds were scarcely audible in this song of woe. "Come back, oh, come back!" Only rarely did she interrupt her song to fly down to the meadows and comb the whole district in search of Greengold and the others. "Kiaw," she called, this time in earnest, not in song—a subtle difference. As time went on, these outbursts of longing became fewer and she spent most of her time perched on the weathercock of our clock-tower, consoling herself with the subdued bars of her song. And here, mourning for her lost love, Greengold, with a veritable

> Green and yellow melancholy,
> She sat like Patience on a monument,
> Smiling at grief.

That is how Redgold saved the colony. For, though I am not given to over-sentimental pity of animals, I was compelled by her grief and her unceasing lament from the rooftop to raise another batch of young jackdaws that spring and to start the jackdaw colony again on our house in Altenberg. For her sake, I reared four young birds and, as soon as they could fly, I put them in the aviary with Redgold. But alas, in my hurry and in my preoccupation with other affairs, I overlooked the fact that there was another large hole in the wire of the cage, and before they had got accustomed to Redgold all the four young jackdaws escaped. Holding closely together and vainly seeking leadership amongst themselves, as I have already described, they circled higher and higher and finally landed away upon the hillside, far from the house and in the midst of a thick beech covert. There I could not approach them, and as the birds were not trained to respond to my call, I had little hope of ever seeing them again. Of course, Redgold could have recovered them with "Kiaw" calls. Old "consuls" of a colony take care of all younger members that are about to stray, but Redgold did not consider the four youngsters as colony-members, since they had been in her company for little more than half a day. Things certainly looked at their blackest, when, all at once, my despair gave place to a brilliant idea: I climbed into the loft and the next moment came crawling out again. Under my arm I bore the huge black-and-gold flag which had flown, to celebrate many birthdays of the late Emperor Francis Joseph I, from the top of my father's house. And high up on the battlements of the roof, hard by the lightning conductor, I now frantically waved this political anachronism. What

was my purpose? I was trying, with this "scarecrow", to drive Redgold so high into the air that the youngsters in the copse would sight her and begin to call. Then, I hoped, the old bird would answer with a "Kiaw" reaction and so bring about the prodigals' return. Redgold circled high, but still not high enough. I let out one Red Indian whoop after the other and waved Francis Joseph's banner like a madman! In the village street, a crowd began to collect. I postponed the explanation of my doings till later, and waved and whooped further. Redgold soared a couple of yards higher, and then—a young jackdaw called from the hillside. I ceased my flag-flying, and, panting, looked above me where the old jackdaw was circling. And, by all the bird-headed gods of Egypt, the beat of her wings had taken on a new vigour, she was scaling higher and higher and now she set her course in the direction of the forest. "Kiaw," she called, "Kiaw, Kiaw"—"Come back, come back!" I wound up the flag with alacrity, and was gone through the trap-door in an instant.

Ten minutes later all four truants were home, in company with Redgold. She was just as tired as I was. But from that day on, she tended those young birds most solicitously and never let them fly away again. These five birds were the nucleus from which a well-populated colony soon developed. At its head stood a female, Redgold. The great disparity in age between her and the other birds gave her even more "authority" than is customary with the despot of a colony. In her ability to hold the flock together, Redgold surpassed all other rulers that my settlement had previously known. Faithfully she herded the young birds, mothering them tenderly because she herself had no children left.

It would be romantic to close here the life-story of the jackdaw Redgold: the altruistic widow safeguarding the prosperity of the flock . . . that indeed would sound no harsh final chord. What really took place makes such an improbable happy ending that I scarcely dare to relate it. It was three years after the great jackdaw catastrophe, and a windy, sun-kissed morning in early spring. Such days are specially favourable for bird migration, and one flock of jackdaws and crows after the other was blowing across the skies. Suddenly a wingless torpedo-shaped projectile separated itself from a group and swooped with gathering speed into the depths below. Hard above our roof-top it stopped its fall with a light swinging movement and landed weightlessly on the weathercock. There sat a big and beautiful jackdaw with blue-black shining wings and a silky nape that gleamed almost white. And Redgold the queen, Redgold the despot surrendered without a thrust. The imperious virago became suddenly a shy, subdued maiden that shook her tail and quivered her wings with all the coyness of a jackdaw bride.

A few hours after the arrival of the newcomer the two were as one mind with but a single thought, and behaved exactly like a long-wedded pair. It was interesting that this big male experienced little or no opposition from the other jackdaws. His recognition as despot by the erstwhile ruler seemed to stamp him as "Number One" in the eyes of all members of the colony.

I have no irrefutable scientific proof that this gorgeous jackdaw male was Greengold, the lost spouse of Redgold the despot. The coloured celluloid rings were broken and gone; Redgold, too, had lost hers long ago. But the new arrival was undoubtedly a member of the former colony; this was proved by his tameness and the readiness with which he entered the interior of the loft. Wild jackdaws which had joined our colony always behaved quite differently. This bird was definitely one of the four or five eldest—the "consuls"—of the first colony. Still, I believe, and hope, that the old rake was no other than Greengold himself. The reunited pair have since hatched and reared many a further brood of hopeful young jackdaws. To-day there are more jackdaws than nesting-holes in Altenberg. In every wall niche, in every chimney is a nest.

Long before the last war, my father, in his autobiography, wrote of the Altenberg jackdaws: "Flocks of these birds fly, particularly towards evening, round the high gables, and communicate by means of penetrating cries. Sometimes I am convinced that I understand them: as perennial retainers, true to our home, we will fly round this, our eyrie, as long as one stone stands upon the other to afford us protection".

The perennial retainers! It is perhaps this quality of the jackdaws which gives them a place in our affection. When in autumn, or even on mild winter days, they tune in their spring songs, when they play their daring game with the raging storm, they touch within me that same chord which sounds when I hear a wren singing on a clear frosty day or when I see an evergreen in snow. They suffuse me with that feeling of hope and fortitude for which the Christmas tree has become a symbol.

Jock has been gone a long time, the victim of an uncertain fate. Redgold was shot in her old age by a kind neighbour with an airgun. I found her dead in the garden. But the jackdaw colony in Altenberg still thrives. Jackdaws fly round our house, steering those courses which Jock was the first to discover, and using the same up-currents that Jock first exploited to gain height. They follow loyally all the traditions which reigned in the first colony, and which were transmitted to the present one through the medium of Red-gold.

How thankful I should be to fate, if I could find but one path which, generations after me, might be trodden by fellow-members of my species. And how infinitely grateful I should be, if, in my life's work, I could find one small "up-current" which might lift some other scientist to a point from which he could see a little further than I do.

12. MORALS AND WEAPONS

They that have power to hurt and will do none,
That do not do the thing they most do show. . . .

SHAKESPEARE, *Sonnets*

IT is early one Sunday morning at the beginning of March, when Easter is already in the air, and we are taking a walk in the Vienna forest whose wooded slopes of tall beeches can be equalled in beauty by few and surpassed by none. We approach a forest glade. The tall smooth trunks of the beeches soon give place to the Hornbeam which are clothed from top to bottom with pale green foliage. We now tread slowly and more carefully. Before we break through the last bushes and out of cover on to the free expanse of the meadow, we do what all wild animals and all good naturalists, wild boars, leopards, hunters and zoologists would do under similar circumstances: we reconnoitre, seeking, before we leave our cover, to gain from it the advantage which it can offer alike to hunter and hunted—namely, to see without being seen.

Here, too, this age-old strategy proves beneficial. We do actually see someone who is not yet aware of our presence, as the wind is blowing away from him in our direction: in the middle of the clearing sits a large fat hare. He is sitting with his back to us, making a big V with his ears, and is watching intently something on the opposite edge of the meadow. From this point, a second and equally large hare emerges and with slow, dignified hops makes his way towards the first one. There follows a measured encounter, not unlike the meeting of two strange dogs. This cautious mutual taking stock soon develops into sparring. The two hares chase each other round, head to tail, in minute circles. This giddy rotating continues for quite a long time. Then suddenly, their pent-up energies burst forth into a battle royal. It is just like the outbreak of war, and happens at the very moment when the long mutual threatening of the hostile parties has forced one to the conclusion that neither dares to make a definite move. Facing each other, the hares rear up on their hind legs and, straining to their full height, drum furiously at each other with their forepads. Now they clash in flying leaps and, at last, to the accompaniment of squeals and grunts, they discharge a volley of lightning kicks, so rapidly that only a slow-motion camera could help us to discern the mechanism of these hostilities. Now, for the time being, they have had enough, and they recommence their circling, this time much faster than before; then follows a fresh, more embittered bout. So engrossed are the two champions that there is nothing to prevent myself and my little daughter from tiptoeing nearer, although that venture cannot be accomplished in silence. Any normal and sensible hare would have heard us long ago, but this is March, and March Hares are mad! The whole boxing match looks so comical that my little daughter, in spite of her iron upbringing in the matter of silence when watching animals, cannot restrain a chuckle. That is too much even for March Hares—two flashes in two different directions and the meadow is empty, while over the battlefield floats a fistful of fluff, light as a thistledown.

It is not only funny, it is almost touching, this duel of the unarmed, this raging fury of the meek in heart. But are these creatures really so meek? Have they really got softer hearts than those of the fierce beasts of prey? If, in a zoo, you ever watched two lions, wolves or eagles in conflict, then, in all probability, you did not feel like laughing. And yet these sovereigns come off no worse than the harmless hares. Most people have the habit of judging carnivorous and herbivorous animals by quite inapplicable moral criteria. Even in fairy tales, animals are portrayed as being a community comparable to that of mankind, as though all species of animals were beings of one and the same family, as human beings are. For this reason, the average person tends to regard the animal that kills animals in the same light as he would the man that kills his own kind. He does not judge the fox that kills a hare by the same standard as the hunter who shoots one for precisely the same reason, but with that severe censure that he would apply to the gamekeeper who made a practice of shooting farmers and frying them for supper! The "wicked" beast of prey is branded as a murderer, although the fox's hunting is quite as legitimate and a great deal more necessary to his existence than is that of the gamekeeper, yet nobody regards the latter's "bag" as his prey, and only one author, whose own standards were indicted by the severest moral criticism, has dared to dub the fox-hunter "the unspeakable in pursuit of the uneatable"! In their dealing with members of their own species, the beasts and birds of prey are far more restrained than many of the "harmless" vegetarians.

Still more harmless than a battle of hares appears the fight between turtle- or ring-doves. The gentle pecking of the frail bill, the light flick of the fragile wing seems, to the uninitiated, more like a caress than an attack. Some time ago I decided to breed a cross between the African blond ring-dove and our own indigenous somewhat frailer turtle-dove, and with this object I put a tame, home-reared male turtle-dove and a female ring-dove together in a roomy cage. I did not take their original scrapping seriously. How could these paragons of love and virtue dream of harming one another? I left them in their cage and went to Vienna. When I returned, the next day, a horrible sight met my eyes.

The turtle-dove lay on the floor of the cage; the top of his head and neck, as also the whole length of his back, were not only plucked bare of feathers, but so flayed as to form a single wound dripping with blood. In the middle of this gory surface, like an eagle on his prey, stood the second harbinger of peace. Wearing that dreamy facial expression that so appeals to our sentimental observer, this charming lady pecked mercilessly with her silver bill in the wounds of her prostrated mate. When the latter gathered his last resources in a final effort to escape, she set on him again, struck him to the floor with a light clap of her wing and continued with her slow pitiless work of destruction. Without my interference she would undoubtedly have finished him off, in spite of the fact that she was already so tired that she could hardly keep her eyes open. Only in two other instances have I seen similar horrible lacerations inflicted on their own kind by vertebrates: once as an observer of the embittered fights of cichlid fishes who sometimes actually skin each other, and again as a field surgeon, in the late war, where the highest of all vertebrates perpetrated mass mutilations on members of his own species. But to return to our "harmless" vegetarians. The battle of the hares which we witnessed in the forest clearing would have ended in quite as horrible a carnage as that of the doves, had it taken place in the confines of a cage where the vanquished could not flee the victor.

If this is the extent of the injuries meted out to their own kind by our gentle doves and hares, how much greater must be the havoc wrought amongst themselves by those beasts to whom nature has relegated the strongest weapons with which to kill their prey? One would certainly think so, were it not that a good naturalist should always check by observation even the most obvious-seeming inferences before he accepts them as truth. Let us examine that symbol of cruelty and voraciousness, the wolf. How do these creatures conduct themselves in their dealings with members of their own species? At Whipsnade, that zoological country paradise, there lives a pack of timber wolves. From the fence of a pine-wood of enviable dimensions we can watch their daily round in an environment not so very far removed from conditions of real freedom. To begin with, we wonder why the antics of the many woolly, fat-pawed cubs have not led them to destruction long ago. The efforts of one ungainly little chap to break into a gallop have landed him in a very different situation from that which he intended. He stumbles and bumps heavily into a wicked-looking old sinner. Strangely enough, the latter does not seem to notice it, he does not even growl. But now we hear the rumble of battle sounds! They are low, but more ominous than those of a dog-fight. We were watching the cubs and have therefore only become aware of this adult fight now that it is already in full swing.

An enormous old timber wolf and a rather weaker, obviously younger one are the opposing champions, and they are moving in circles round each other, exhibiting admirable "footwork". At the same time, the bared fangs flash in such a rapid exchange of snaps that the eye can scarcely follow them. So far, nothing has really happened. The jaws of one wolf close on the gleaming white teeth of the other, who is on the alert and wards off the attack. Only the lips have received one or two minor injuries. The younger wolf is gradually being forced backwards. It dawns upon us that the older one is purposely manœuvring him towards the fence. We await with breathless anticipation what will happen when he "goes to the wall". Now he strikes the wire netting, stumbles . . . and the old one is upon him. And now the incredible happens, just the opposite of what you would expect. The furious whirling of the grey bodies has come to a sudden standstill. Shoulder to shoulder they stand, pressed against each other in a stiff and strained attitude, both heads now facing in the same direction. Both wolves are growling angrily, the elder in a deep bass, the younger in higher tones, suggestive of the fear that underlies his threat. But notice carefully the position of the two opponents; the older wolf has his muzzle close, very close against the neck of the younger, and the latter holds away his head, offering unprotected to his enemy the bend of his neck, the most vulnerable part of his whole body! Less than an inch from the tensed neck-muscles, where the jugular vein lies immediately beneath the skin, gleam the fangs of his antagonist from beneath the wickedly retracted lips. Whereas, during the thick of the fight, both wolves were intent on keeping only their teeth, the one invulnerable part of the body, in opposition to each other, it now appears that the discomfited fighter proffers intentionally that part of his anatomy to which a bite must assuredly prove fatal. Appearances are notoriously deceptive, but in his case, surprisingly, they are not!

This same scene can be watched any time wherever street-mongrels are to be found. I cited wolves as my first example because they illustrate my point more impressively than the all-too-familiar domestic dog. Two adult male dogs meet in the street. Stiff-legged, with tails erect and hair on end, they pace towards each other. The nearer they approach, the stiffer, higher and more ruffled they appear, their advance becomes slower and slower. Unlike fighting cocks they do not make their encounter head to head, front against front, but make as though to pass each other, only stopping when they stand at last flank to flank, head to tail, in close juxtaposition. Then a strict ceremonial demands that each should sniff the hind regions of the other. Should one of the dogs be overcome with fear at this juncture, down goes his tail between his legs and he jumps with a quick, flexible twist, wheeling at an angle of 180 degrees, thus modestly retracting his former offer to be smelt. Should the two dogs remain in an attitude of self-display, carrying their tails as rigid as standards, then the sniffing process may be of a long-protracted nature. All may be solved amicably and there is still the chance that first one tail and then the other may begin to wag with small but rapidly increasing beats, and then this nerve-racking situation may develop into nothing worse than a cheerful canine romp. Failing this solution the situation becomes more and more tense, noses begin to wrinkle and to turn up with a vile, brutal expression, lips begin to curl, exposing the fangs on the side nearer the opponent. Then the animals scratch the earth angrily with their hind feet, deep growls rise from their

chests, and, in the next moment, they fall upon each other with loud piercing yells.

But to return to our wolves, whom we left in a situation of acute tension. This was not a piece of inartistic narrative on my part, since the strained situation may continue for a great length of time which is minutes to the observer but very probably seems hours to the losing wolf. Every second you expect violence and await with bated breath the moment when the winner's teeth will rip the jugular vein of the loser. But your fears are groundless, for it will not happen. In this particular situation, the victor will definitely not close on his less fortunate rival. You can see that he would like to, but he just cannot! A dog or wolf that offers its neck to its adversary in this way will never be bitten seriously. The other growls and grumbles, snaps with his teeth in the empty air and even carries out, without delivering so much as a bite, the movement of shaking something to death in the empty air. However, this strange inhibition from biting persists only so long as the defeated dog or wolf maintains his attitude of humility. Since the fight is stopped so suddenly by this action, the victor frequently finds himself straddling his vanquished foe in anything but a comfortable position. So to remain with his muzzle applied to the neck of the "under-dog" soon becomes tedious for the champion, and, seeing that he cannot bite anyway, he soon withdraws. Upon this, the under-dog may hastily attempt to put distance between himself and his superior. But he is not usually successful in this, for as soon as he abandons his rigid attitude of submission the other again falls upon him like a thunderbolt and the victim must again freeze into his former posture. It seems as if the victor is only waiting for the moment when the other will relinquish his submissive attitude, thereby enabling him to give vent to his urgent desire to bite. But, luckily for the "under-dog", the top-dog at the close of the fight is overcome by the pressing need to leave his trademark on the battlefield, to designate it as his personal property—in other words, he must lift his leg against the nearest upright object. This right-of-possession ceremony is usually taken advantage of by the under-dog to make himself scarce.

By this commonplace observation, we are here, as so often, made conscious of a problem which is actual in our daily life and which confronts us on all sides in the most various forms. Social inhibitions of this kind are not rare but so frequent that we take them for granted and do not stop to think about them. An old German proverb says that one crow will not peck out the eye of another, and for once the proverb is right. A tame crow or raven will no more think of pecking at your eye than he will at that of one of his own kind. Often when Roah, my tame raven, was sitting on my arm, I purposely put my face so near to his bill that my open eye came close to its wickedly curved point. Then Roah did something positively touching. With a nervous, worried movement he withdrew his beak from my eye, just as a father who is shaving will hold back his razor-blade from the inquisitive fingers of his tiny daughter. Only in one particular connection did Roah ever approach my eye with his bill during this facial grooming. Many of the higher, social birds and mammals—above all monkeys—will groom the skin of a fellow-member of their species in those parts of his body to which he himself cannot obtain access. In birds, it is particularly the head and the region of the eyes which are dependent on the attentions of a fellow. In my description of the jackdaw I have already spoken of the gestures with which these birds invite one another to preen their head feathers. When, with half-shut eyes, I held my head sideways towards Roah, just as corvine birds do to each other, he understood this movement in spite of the fact that I have no head feathers to ruffle, and at once began to groom me. While doing so, he never pinched my skin, for the epidermis of birds is delicate and would not stand such rough treatment. With wonderful precision, he submitted every attainable hair to a dry-cleaning process by drawing it separately through his bill. He worked with the same intensive concentration that distinguishes the "lousing" monkey and the operating surgeon. This is not meant as a joke: the social grooming of monkeys, and particularly of anthropoid apes, has not the object of catching vermin—these animals usually have none—and is not limited to the cleaning of the skin, but serves also more remarkable operations, for instance the dexterous removal of thorns and even the squeezing-out of small carbuncles.

The manipulations of the dangerous-looking corvine beak round the open eye of a man naturally appear ominous, and of course I was always receiving warnings from onlookers at this procedure: "You never know—a raven is a raven—" and similar words of wisdom. I used to respond with the paradoxical observation that the warner was for me potentially more dangerous than the raven. It has often happened that people have been shot dead by madmen who have masked their condition with the cunning and pretence typical of such cases. There was always a possibility, though admittedly a very small one, that our kind adviser might be afflicted with such a disease. But a sudden and unpredictable loss of the eye-pecking inhibition in a healthy, mature raven is more unlikely by far than an attack by a well-meaning friend.

Why has the dog the inhibition against biting his fellow's neck? Why has the raven an inhibition against pecking the eye of his friend? Why has the ring-dove no such "insurance" against murder? A really comprehensive answer to these questions is almost impossible. It would certainly involve a *historical* explanation of the process by which these inhibitions have been developed in the course of evolution.

There is no doubt that they have arisen side by side with the development of the dangerous weapons of the beast of prey. However, it is perfectly obvious why these inhibitions are necessary to all weapon-bearing animals. Should the raven peck, without compunction, at the eye of his nest-mate, his wife or his young, in the same way as he pecks at any other moving and glittering object, there would, by now, be no more ravens in the world. Should a dog or wolf unrestrainedly and unaccountably bite the neck of his pack-mates and actually execute the movement of shaking them to death, then his species also would certainly be exterminated within a short space of time.

The ring-dove does not require such an inhibition since it can only inflict injury to a much lesser degree, while its ability to flee is so well developed that it suffices to protect the bird even against enemies equipped with vastly better weapons. Only under the unnatural conditions of close confinement which deprive the losing dove of the possibility of flight does it become apparent that the ring-dove has no inhibitions which prevent it from injuring or even torturing its own kind. Many other "harmless" herbivores prove themselves just as unscrupulous when they are kept in narrow captivity. One of the most disgusting, ruthless and bloodthirsty murderers is an animal which is generally considered as being second only to the dove in the proverbial gentleness of its nature, namely the roe-deer. The roe-buck is about the most malevolent beast I know and is possessed, into the bargain, of a weapon, its antlers, which it shows mighty little restraint in putting into use. The species can "afford" this lack of control since the fleeing capacity even of the weakest doe is enough to deliver it from the strongest buck. Only in very large paddocks can the roe-buck be kept with females of his own kind. In smaller enclosures, sooner or later he will drive his fellows, females and young ones included, into a corner and gore them to death. The only "insurance against murder" which the roe-deer possesses is based on the fact that the onslaught of the attacking buck proceeds relatively slowly. He does not rush with lowered head at his adversary as, for example, a ram would do, but he approaches quite slowly, cautiously feeling with his antlers for those of his opponent. Only when the antlers are interlocked and the buck feels firm resistance does he thrust with deadly earnest. According to the statistics given by W. T. Hornaday, the former director of the New York Zoo, tame deer cause yearly more serious accidents than captive lions and tigers, chiefly because an uninitiated person does not recognize the slow approach of the buck as an earnest attack, even when the animal's antlers have come dangerously near. Suddenly there follows, thrust upon thrust, the amazingly strong stabbing movement of the sharp weapon, and you will be lucky if you have time enough to get a good grip on the aggressor's antlers. Now there follows a wrestling-match in which the sweat pours and the hands drip blood, and in which even a very strong man can hardly obtain mastery over the roe-buck unless he succeeds in getting to the side of the beast and bending his neck backwards. Of course, one is ashamed to call for help—until one has the point of an antler in one's body! So take my advice and if a charming, tame roe-buck comes playfully towards you, with a characteristic prancing step and flourishing his antlers gracefully, hit him, with your walking-stick, a stone or the bare fist, as hard as you can, on the side of his nose, before he can apply his antlers to your person.

And now, honestly judged: who is really a "good" animal—my friend Roah to whose social inhibitions I could trust the light of my eyes, or the gentle ring-dove that in hours of hard work nearly succeeded in torturing its mate to death? Who is a "wicked" animal, the roe-buck who will slit the bellies even of females and young of his own kind if they are unable to escape him, or the wolf who cannot bite his hated enemy if the latter appeals to his mercy?

Now let us turn our mind to another question. Wherein consists the essence of all the gestures of submission by which a bird or animal of a social species can appeal to the inhibitions of its superior? We have just seen, in the wolf, that the defeated animal actually facilitates his own destruction by offering to the victor those very parts of his body which he was most anxious to shield as long as the battle was raging. All submissive attitudes with which we are so far familiar, in social animals, are based on the same principle: The supplicant always offers to his adversary the most vulnerable part of his body, or, to be more exact, that part *against which every killing attack is inevitably directed*! In most birds this area is the base of the skull. If one jackdaw wants to show submission to another, he squats back on his hocks, turns away his head, at the same time drawing in his bill to make the nape of his neck bulge, and, leaning towards his superior, seems to invite him to peck at the fatal spot. Seagulls and herons present to their superior the top of their head, stretching their neck forward horizontally, low over

the ground, also a position which makes the supplicant particularly defenceless.

With many gallinaceous birds, the fights of the males commonly end by one of the combatants being thrown to the ground, held down and then scalped as in the manner described in the ring-dove. Only one species shows mercy in this case, namely the turkey: and this one only does so in response to a specific submissive gesture which serves to forestall the intent of the attack. If a turkey-cock has had more than his share of the wild and grotesque wrestling-match in which these birds indulge, he lays himself with

outstretched neck upon the ground. Whereupon the victor behaves exactly as a wolf or dog in the same situation—that is to say, he evidently *wants* to peck and kick at the prostrated enemy, but simply cannot: he would if he could but he can't! So, still in threatening attitude, he walks round and round his prostrated rival, making tentative passes at him, but leaving him untouched.

This reaction—though certainly propitious for the turkey species—can cause a tragedy if a turkey comes to blows with a peacock, a thing which not infrequently happens in captivity, since these species are closely enough related to "appreciate" respectively their mutual manifestations of virility. In spite of greater strength and weight the turkey nearly always loses the match, for the peacock flies better and has a different fighting technique. While the red-brown American is muscling himself up for the wrestling-match, the blue East Indian has already flown above him and struck at him with his sharply pointed spurs. The turkey justifiably considers this infringement of his fighting code as unfair and, although he is still in possession of his full strength, he throws in the sponge and lays himself down in the manner described above. And a ghastly thing happens: the peacock does not "understand" this submissive gesture of the turkey —that is to say, it elicits no inhibition of his fighting drives. He pecks and kicks further at the helpless turkey, who, if nobody comes to his rescue, is doomed, for the more pecks and blows he receives, the more certainly are his escape reactions blocked by the psycho-physiological mechanism of the submissive attitude. It does not and cannot occur to him to jump up and run away.

The fact that many birds have developed special "signal organs" for eliciting this type of social inhibition, shows convincingly the blind instinctive nature and the great evolutionary age of these submissive gestures. The young of the water-rail, for example, have a bare red patch at the back of their head which, as they present it meaningly to an older and stronger fellow, takes on a deep red colour. Whether, in higher animals and man, social inhibitions of this kind are equally mechanical, need not for the moment enter into our consideration. Whatever may be the reasons that prevent the dominant individual from injuring the submissive one, whether he is prevented from doing so by a simple and purely mechanical reflex process or by a highly philosophical moral standard, is immaterial to the practical issue. The essential behaviour of the submissive as well as of the dominant partner remains the same: the humbled creature suddenly seems to lose his objections to being injured and removes all obstacles from the path of the killer, and it would seem that the very removal of these outer obstacles raises an insurmountable inner obstruction in the central nervous system of the aggressor.

And what is a human appeal for mercy after all? Is it so very different from what we have just described? The Homeric warrior who wishes to yield and plead mercy, discards helmet and shield, falls on his knees and inclines his head: a set of actions which should make it easier for the enemy to kill, but, in reality, hinders him from doing so. As Shakespeare makes Nestor say of Hector:

> Thou hast hung thy advanced sword i' the air,
> Not letting it decline on the declined.

Even to-day we have retained many symbols of such submissive attitudes in a number of our gestures of courtesy: bowing, removal of the hat, and presenting arms in military ceremonial. If we are to believe the ancient epics, an appeal

to mercy does not seem to have raised an "inner obstruction" which was entirely insurmountable. Homer's heroes were certainly not as soft-hearted as the wolves of Whipsnade! In any case, the poet cites numerous instances where the supplicant was slaughtered with or without compunction. The Norse heroic sagas bring us many examples of similar failures of the submissive gesture, and it was not till the era of knight-errantry that it was no longer considered "sporting" to kill a man who begged for mercy. The Christian knight is the first who, for reasons of traditional and religious morals, is as chivalrous as is the wolf from the depth of his natural impulses and inhibitions. What a strange paradox!

Of course, the innate, instinctive, fixed inhibitions that prevent an animal from using his weapons indiscriminately against his own kind are only a functional analogy, at the most a slight foreshadowing, a genealogical predecessor of the social morals of man. The worker in comparative ethology does well to be very careful in applying moral criteria to animal behaviour. But here I must myself own to harbouring sentimental feelings: I think it a truly magnificent thing that one wolf finds himself unable to bite the proffered neck of the other, but still more so that the other relies upon him for this amazing restraint. Mankind can learn a lesson from this, from the animal that Dante calls "la bestia senza pace". I at least have extracted from it a new and deeper understanding of a wonderful and often misunderstood saying from the Gospel which hitherto had only awakened in me feelings of strong opposition: "And unto him that smiteth thee on the one cheek offer also the other" (St Luke VI, 29). A wolf has enlightened me: not so that your enemy may strike you again do you turn the other cheek toward him, but to make him unable to do it.

When, in the course of its evolution, a species of animals develops a weapon which may destroy a fellow-member at one blow, then, in order to survive, it must develop, along with the weapon, a social inhibition to prevent a usage which could endanger the existence of the species. Among the predatory animals there are only a few which lead so solitary a life that they can, in general, forgo such restraint. They come together only at the mating season when the sexual impulse outweighs all others, including that of aggression. Such unsociable hermits are the polar bear and the jaguar; owing to the absence of these social inhibitions, animals of these species, when kept together in zoos, hold a sorry record for murdering their own kind. The system of special inherited impulses and inhibitions, together with the weapons with which a social species is provided by nature, form a complex which is carefully computed and self-regulating. All living beings have received their weapons through the same process of evolution that moulded their impulses and inhibitions; for the structural plan of the body and the system of behaviour of a species are parts of the same whole.

> If such be Nature's holy plan,
> Have I not reason to lament
> What man has made of man?

Wordsworth is right: there is only one being in possession of weapons which do not grow on his body and of whose working plan, therefore, the instincts of his species know

nothing and in the usage of which he has no correspondingly adequate inhibition. That being is man. With unarrested growth his weapons increase in monstrousness, multiplying horribly within a few decades. But innate impulses and inhibitions, like bodily structures, need time for their development—time on a scale in which geologists and astronomers are accustomed to calculate, and not historians. We did not receive our weapons from nature. We made them ourselves, of our own free will. Which is going to be easier for us in the future, the production of the weapons or the engendering of the feeling of responsibility that should go along with them, the inhibitions without which our race must perish by virtue of its own creations? We must build up these inhibitions purposefully, for we cannot rely upon our instincts. Fourteen years ago, in November 1935, I concluded an article on "Morals and Weapons of Animals", which appeared in a Viennese journal, with the words, "The day will come when two warring factions will be faced with the possibility of each wiping the other out completely. The day may come when the whole of mankind is divided into two such opposing camps. Shall we then behave like doves or like wolves? The fate of mankind will be settled by the answer to this question." We may well be apprehensive.

INDEX